T0122847

Praise for

FIELD GUIDE to FRESHWATER FISHES of VIRGINIA

"With jaw-dropping color illustrations and descriptions that make you feel like you know these fishes personally, *Field Guide to Freshwater Fishes of Virginia* has everything but a scratch-and-sniff section. Those who love these fishes will find it impossible to get a better representation of the aquatic life of Virginia than they will find in this glorious new book."

—Prosanta Chakrabarty, Associate Professor and Curator of Fishes, Louisiana State University

"An excellent distillation of information on the fishes of Virginia that is truly usable as a field guide. High-quality color illustrations by Val Kells and Joseph Tomelleri and informative range maps make for visually informative species accounts. It's a book you'll want to have streamside in Virginia."

—Bruce W. Stallsmith, Professor, Biological Sciences, University of Alabama in Huntsville

"This beautiful and succinct book has some of the best illustrations out there right now. Strongly recommended. All federal biologists and academics in aquatic sciences should buy a copy."

—Brooks M. Burr, Emeritus Professor, Southern Illinois University at Carbondale

"*Field Guide to Freshwater Fishes of Virginia* is an absolute home run. Readers learn about the major watersheds in the state, the different habitats stretching from the mountains to the sea, the anatomy of the fishes that inhabit these waters, and finally, how to identify 225 species. The text is easy to read and often funny, and the illustrations are outstanding. Generations to come are indebted to the folks who put so much time and energy into getting this guide just right."

—Charles Gowan, Paul H. Wornom Professor of Biology, Randolph-Macon College

"This guide is a stunning portrayal of the high diversity and colorful beauty of the freshwater fishes of Virginia. It is essential for identifying the species and summarizes their biology and geography. Conservation needs within the fauna are pervasively recognized."

—Bob Jenkins, Professor Emeritus, Roanoke College, and coauthor of Freshwater Fishes of Virginia

"A fantastic new field guide that will be welcomed by many, and not just by those who live in Virginia. It is easily carried into the field, with a wealth of information in its pages. Val Kells's and Joseph Tomelleri's illustrations are top notch."

—Fritz Rohde, Fisheries Biologist, NOAA, coauthor of Freshwater Fishes of South Carolina, *editor of* American Currents, *and President of the North American Native Fishes Association*

"The combination of stellar artwork, comparative distribution maps, and informative species descriptions will aid both hobbyists and scientific experts in the identification of Virginia's freshwater fishes. The exceptional detail of this field guide, in concert with exemplary photographs, bring to life the amazing ichthyofaunal diversity of this beautiful region."

—Martin C. Arostegui, University of Washington, IGFA Representative and Lifetime Achievement Award Winner

"*Field Guide to Freshwater Fishes of Virginia* is a must-have for avid anglers, aquatic scientists, and anyone who likes to play around in Virginia's lakes and streams. The illustrations are incredible, the descriptions are easy to understand, and the range maps make species identification a breeze. I wish I'd had this book 30 years ago when I was an amateur biologist, and will certainly use it as a professional now."

—J. Wesley Neal, Certified Fisheries Professional, Department of Wildlife, Fisheries & Aquaculture and Extension Professor, Mississippi State University

"An outstanding work for anyone—scientist, student, angler, or naturalist—fascinated by fishes, in Virginia and beyond. Beautiful illustrations and detailed species accounts, along with identification keys, information on management and conservation, and even tips for watching and enjoying fishes make this guide a must-have in your fish reference collection."

—Rob Neumann, Managing Editor and Fishery Scientist, In-Fisherman.com

"*Field Guide to Freshwater Fishes of Virginia* is an ideal resource for anyone wanting to learn about the biology and ecology of Virginia's fishes. I see this guide becoming a trusted information source for volunteers like Virginia Master Naturalists, especially those involved in citizen science studies in freshwater habitats or educating the public about Virginia's vast and fascinating diversity of fishes. Nature enthusiasts, anglers, and even natural resource professionals will also find this guide to be an excellent tool for learning and practicing fish identification. The illustrations and the key to fish families are particularly helpful."

—Michelle D. Prysby, Director, Virginia Master Naturalist Program, Department of Forest Resources and Environmental Conservation, Virginia Tech, Virginia Cooperative Extension

"This much-needed new field guide fills a niche for educators and fish enthusiasts. The diversity of Virginia's freshwaters is beautifully captured in the photographs and full-color illustrations. The descriptions provide clear diagnostics for fish species identification to go along with the detailed illustrations, and the range maps given for each species make it quick and easy to confirm species presence in your area. The introductory chapters provide excellent guidance on how to navigate species identification as well as on the ecology and conservation concerns of our aquatic ecosystems."

—Robert Humston, Director for Environmental Studies and Professor of Biology, Washington and Lee University

"As someone who has spent a lifetime hunting whatever obscure species I could find in whatever venue I could fish—including golf courses and hotel fountains—this book shows me that I haven't even scratched the surface. I'm off to Virginia!"

—Steve Wozniak, 1000fishes.com, IGFA World Record Holder, contributor, Sportfishing Magazine

"*Field Guide to Freshwater Fishes of Virginia* will be a great resource for hobbyists, naturalists, and fisheries professionals alike. It has a convenient and easy to use layout. Similar species being mapped together helps the reader use the location of their captured/observed fish to reach a conclusion on what they have found, without having to compare several maps. This field guide will be a nice complement to older, much more technical volumes, and a nice addition to the libraries of fish enthusiasts."

—Brian Zimmerman, Research Associate, The Ohio State University

"In *Field Guide to Freshwater Fishes of Virginia*, Bugas et al. have brought forth an extraordinarily useful and beautifully illustrated field guide to the rich and diverse fish fauna that inhabit the streams, rivers, and lakes of the state. The authors are biologists who have worked extensively with Virginia's fishes, and it features the works of two of America's most talented fish illustrators, prolific author Val Kells and the widely known Joseph Tomelleri. This guide nicely brings together a combination of these works to provide the users with both the best opportunity for accurate identification of fishes in hand, as well as an appreciation for their beauty."

—Wayne C. Starnes, Research Curator of Fishes, North Carolina Museum of Natural Sciences (retired)

"An eminently successful introduction to the underwater world of the fishes of Virginia. As a lifelong birder, the 'how to watch fish' part of this book has encouraged me to begin my 'life list' of fishes. River drainages, aquatic ecology basics, and a documented basis for fish conservation programs are covered in this handsome volume."

—M. Rupert Cutler, Past President, Defenders of Wildlife, Editor Emeritus, Virginia Wildlife *and* National Wildlife, *and former Assistant Secretary for Natural Resources and Environment, USDA*

FIELD GUIDE *to* FRESHWATER FISHES *of* VIRGINIA

Field Guide *to* Freshwater Fishes *of* Virginia

Paul E. Bugas, Jr.
Corbin D. Hilling
Val Kells
Michael J. Pinder
Derek A. Wheaton
Donald J. Orth

Illustrated by
Val Kells
and Joseph R. Tomelleri

© 2019 Johns Hopkins University Press
Illustrations © 2019 Val Kells
All rights reserved. Published 2019
Printed in China on acid-free paper
4 6 8 9 7 5 3

Johns Hopkins University Press
2715 North Charles Street
Baltimore, Maryland 21218-4363
www.press.jhu.edu

Library of Congress Cataloging-in-Publication Data

Names: Bugas, Paul E., Jr, author.
Title: Field guide to freshwater fishes of Virginia / Paul E. Bugas, Jr., Corbin D. Hilling, Val Kells, Michael J. Pinder, Derek A. Wheaton, Donald J. Orth ; illustrated by Val Kells and Joseph R. Tomelleri.
Description: Baltimore : Johns Hopkins University Press, 2019. | Includes bibliographical references and index.
Identifiers: LCCN 2019004928 | ISBN 9781421433059 (paperback : alk. paper) | ISBN 1421433052 (paperback : alk. paper) | ISBN 9781421433073 (electronic) | ISBN 1421433079 (electronic)
Subjects: LCSH: Freshwater fishes—Virginia—Identification.
Classification: LCC QL628.V55 B84 2019 | DDC 597.176—dc23
LC record available at https://lccn.loc.gov/2019004928

A catalog record for this book is available from the British Library.

Special discounts are available for bulk purchases of this book.
For more information, please contact Special Sales at specialsales@jh.edu

To Bob Jenkins and Noel Burkhead for paving the way
with their authoritative Freshwater Fishes of Virginia.
—PEB, CDH, VK, MJP, DAW, DJO

To Ruth, Ryan, Caitlin, and Michael; my parents, Rosemary and Fritz;
and David A. Griffin.
—PEB

To Lindsey, my parents, David and Sadie, and Eric Kincaid.
—CDH

For my sons, Dave and Drew, and my mother, Joasha.
Live, love, fish!
—VK

To my brother, Ed. Although you're gone, your spirit lives on.
Strong Like Bull.
—MJP

To my parents, Michael and Sharon Wheaton, for always encouraging
my love of the natural world, and the members of NANFA for mentoring
me as a young man and guiding me down the right path:
towards the study of fishes.
—DAW

To Kyler, Adalyn, Cora, Nolan, Bodhi,
and future generations of fish enthusiasts
—DJO

CONTENTS

CONTENTS

ACKNOWLEDGMENTS

We thank many people who contributed to this guide and hope we acknowledge all. Black-and-white illustrations were scanned from original art (Jenkins and Burkhead 1994) by Verner A. Plott, Virginia Tech Digital Imaging Services. Range maps were created by Karen Horodysky, Jay Kapalczynski, and Ed Laube, Virginia Department of Game and Inland Fisheries (VDGIF). Experts reviewed specific accounts; these included Brett Albanese, David Argent, Andy Dolloff, Eric Hilton, Tom Kwak, Zach Martin, John Odenkirk, Steven Powers, Fritz Rohde, Jamie Roberts, Steve Sammons, J. R. Shute, Peggy Shute, Dustin Smith, Wayne Starnes, Jeremy Tiemann, Alan Weaver, and Stuart Welsh.

Color illustrations were provided by Val Kells and Joseph R. Tomelleri. Illustrations from Jenkins and Burkhead were reproduced with permission by ©American Fisheries Society. Illustration of River Redhorse lips was created by Michael J. Pinder. Illustrations of the food web and stream habitats are courtesy of the Missouri Department of Conservation and Texas A&M University Press. Illustrations of sculpin caudal fins by Dave Neely are used with permission by © Copeia. Technical support provided by Tara Craig, Chanz Hopkins, and Valerie F. Orth. Tiffany Gasbarrini, Johns Hopkins University Press, confidently guided this book from proposal to completion, and Debby Bors provided invaluable editorial expertise.

Financial contributions for production were given by Dominion Energy, The Wildlife Foundation of Virginia (in partnership with the VDGIF e Store), Claude Moore Charitable Foundation, North American Native Fishes Association, Virginia Chapter of the American Fisheries Society, Virginia Council of Trout Unlimited, Virginia Tech College of Natural Resources, Virginia Tech Fish and Wildlife Conservation, Virginia Tech Chapter of the American Fisheries Society, and Rivanna Conservation Alliance.

Michael J. Pinder and Paul E. Bugas, Jr., acknowledge the support of their employer, VDGIF. Corbin D. Hilling was supported by a Virginia Sea Grant Fellowship and Virginia Tech Graduate Teaching Assistantship. Derek A. Wheaton acknowledges the support of his employer, Conservation Fisheries, Inc. The participation of Donald J. Orth was supported in part by the US Department of Agriculture through the National Institute of Food and Agriculture Program and Virginia Tech University.

The following people provided color photos for the guide: Brian Borkholder, Paul E. Bugas, Jr., Louise Finger, Sean Landsman, Tim Lane, Christine Lisle, Megan Marchetti, Donald J. Orth, Michael J. Pinder, Michael St. Germain, Isaac Szabo, and Derek A. Wheaton. Thank you, all!

INTRODUCTION

HOW TO USE THIS GUIDE

The *Field Guide to Freshwater Fishes of Virginia* is designed to easily identify species and common hybrids of Virginia's freshwater fishes. This field guide will fit the needs of a diverse audience, including fishers, aquarists, educators, students, divers, snorkelers, naturalists, novices, and scientists. It is divided into sections, beginning with introductory sections that include anatomy, a dichotomous key to families, and information about fish fauna, catching, watching, fish keeping, and management and conservation. This is followed by up-to-date family and species descriptions.

The reader should begin by becoming familiar with basic fish anatomy and terms. When encountering an unfamiliar fish, one can begin by using the key to families. With practice, one will quickly learn how to narrow a species down to a single family. Families are organized in the order in which they evolved; the more generalized fish families appear first, and more specialized fish families are later in the field guide.

Jenkins and Burkhead (1994) provided in-depth descriptions in *Freshwater Fishes of Virginia*, which is targeted toward readers with more specialized training. We follow their delineations of fishes and provide updated names and ranges of all fresh- and brackish-water fishes of Virginia. We also cover some species that are primarily estuarine, mention species that are extirpated, and include some that are presently undescribed but are recognized by the Virginia Department of Game and Inland Fisheries.

This book is organized by families and by groups of similar species within those families. Families are presented in phylogenetic order according to *Fishes of the World*, Fourth Edition, 2006. In some cases, species similar in appearance are treated within a single account, with their similarities and differences discussed. Species (or groups of species) are also presented in phylogenetic order.

Species status and scientific names of species presented here follow the California Academy of Sciences Catalog of Fishes, although some taxa have not yet been formally described. Common names of fishes can vary among geographic regions. For instance, the Rock Bass (*Ambloplites rupestris*) is known as "red eye" in many parts of Virginia. The Flathead Catfish (*Pylodictis olivaris*) is known by many names across its range, including "mudcat," "shovelhead catfish," and "pied cat." This can be confusing, especially given that some of these names are used for other fishes. To reduce the confusion of local names, we follow the *Common and Scientific Names of Fishes from the United States, Canada, and Mexico*, Seventh Edition, 2013, for accepted common names.

Each family description provides a short narrative of essential characteristics, history, and size of the family. Each species account provides a color illustration of the fish, scientific name, common name, colloquial names (if appropriate), synonyms, size, abundance, habitat, status, description, reproduction, food, and notes. See sample below.

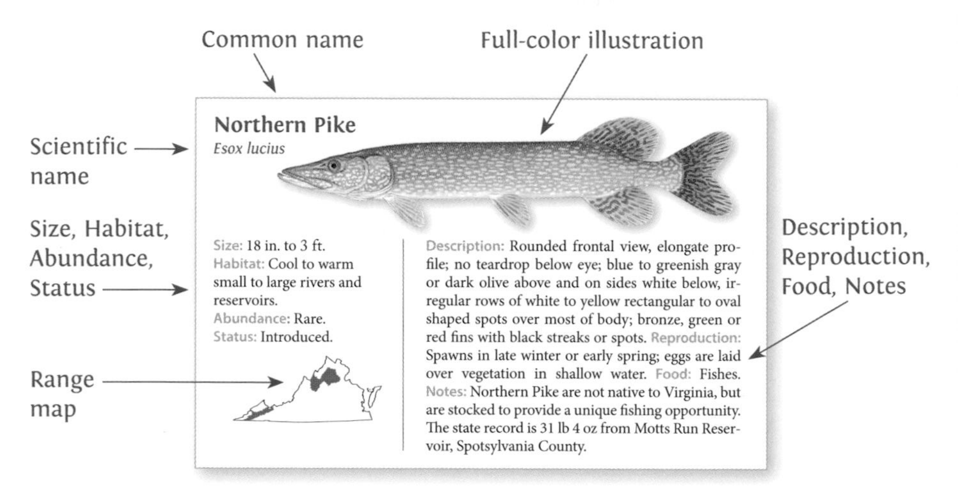

Within the species accounts, readers will find a range of average adult sizes. This is not an absolute minimum or maximum size for the species, but a range in which most of the adults likely fall. Habitat is described as the type of body of water the species is typically found in and its tendencies toward a particular substrate or cover. The abundance of the species is described as abundant, common, uncommon, or rare, within its habitat. The Status section indicates whether a species is native or nonnative. The species accounts also feature detailed descriptions of how members of taxonomic families differ from each other. Body form is described first, followed by coloration. The reproductive behavior of the species is discussed in the Reproduction section and dietary tendencies in the Food section. The Notes section describes any interesting aspects of the species' biology, distribution, or behavior.

Color illustrations provide readers with idealized depictions of representative examples of each species. In cases where similar species were grouped, a representative species is illustrated, and differences between the illustrated species and similar species are provided in the text. This book also includes black-and-white drawings that show the reader anatomical differences between species.

Range maps represent the best available information about where species in this book occur today. The maps show hydrologic units, which represent watersheds, but not specific stream segments, where the fishes occur. However, it is reasonable to expect a species in its appropriate aquatic habitat somewhere in the shaded area.

VIRGINIA'S FRESHWATER FISHES

Introduction

The inland waters of Virginia support an incredible diversity of freshwater fishes. Some are large and well known, such as the bass and trout that are pursued by anglers for sport and food. Many are neither big nor flashy but have interesting habits. Freshwater fishes are just a stream, river, or lake away. We hope by using this guide the reader develops a greater understanding and appreciation for this remarkable group of Virginia's wildlife.

Over 35,000 fish species exist worldwide. Even though only 0.009% of our planet's water is fresh water, 43% of all fish species use fresh water either exclusively or in addition to marine environments. Of the 1,213 freshwater fishes in North America, most are present in the Interior and Appalachian Highlands regions of the southeastern United States. Virginia has 226 freshwater fish species, and Alabama, Tennessee, Georgia, and Mississippi each have even more. Jenkins and Burkhead documented 210 species in 1994. That number has risen since then due to recent discoveries, introductions, and genetic assessments. As we learn more about Virginia's fishes, the total number will likely change.

Fish Diversity

Virginia's fishes are categorized into 25 families (see below). A "family" represents a higher level of taxonomic grouping of species with similar lineages and body characteristics. Familiar game fishes like bass and sunfish, many of which are small and not targeted by anglers, belong to the family Centrachidae. Taxonomically, Virginia's fish families range from the ancestral, jawless Petromyzontidae to the more advanced spiny-rayed family Sciaenidae. While many fish families have only a few members, Cyprinidae and Percidae collectively comprise over half (121) of the total number of species in Virginia.

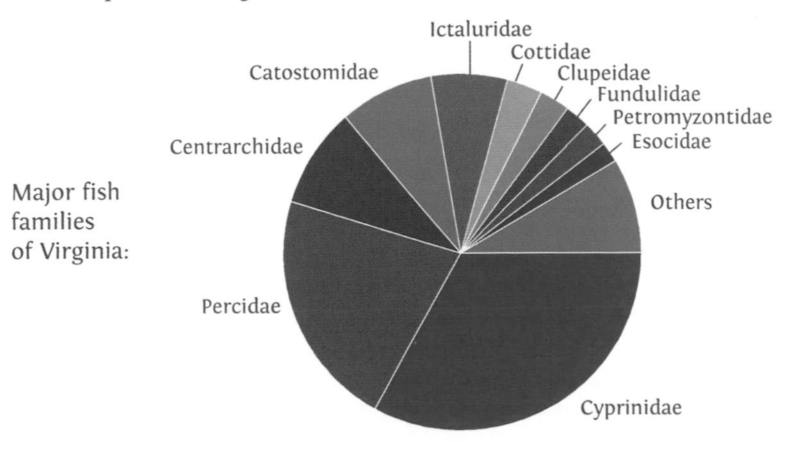

Multiple factors are responsible for Virginia's diversity of fishes. Virginia is situated in both the southern and northern US climates. To the east, Virginia directly adjoins the Chesapeake Bay and Atlantic Ocean. To the west, it features drainages that flow from the Mississippi River into the Gulf of Mexico. As a result, many species are well established in or have their ranges limited to Virginia.

Virginia has a humid, temperate climate averaging 43 inches of rainfall annually. In this relatively wet environment, differences in geology and topography provide diverse aquatic habitats, enabling a wide diversity of fish species to evolve. Historical factors, such as the lack of Pleistocene glaciers, interconnections between river systems, and species dispersal patterns, all contribute to Virginia's rich fish fauna.

Virginia, its adjacent states, and its two major drainage basins: Mississippi (blue) and Atlantic Slope (yellow):

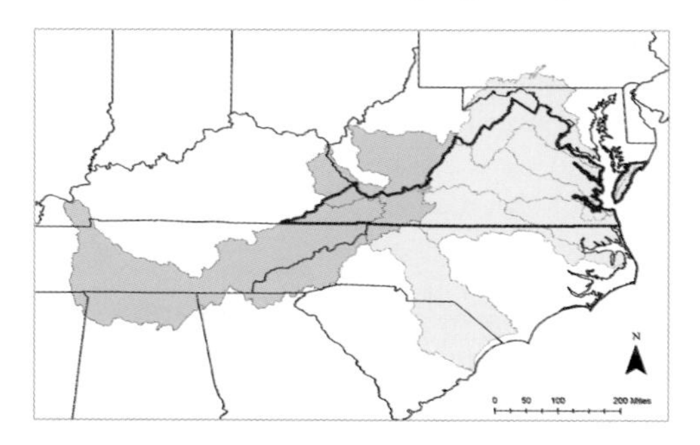

Major physiographic provinces and elevations of Virginia:

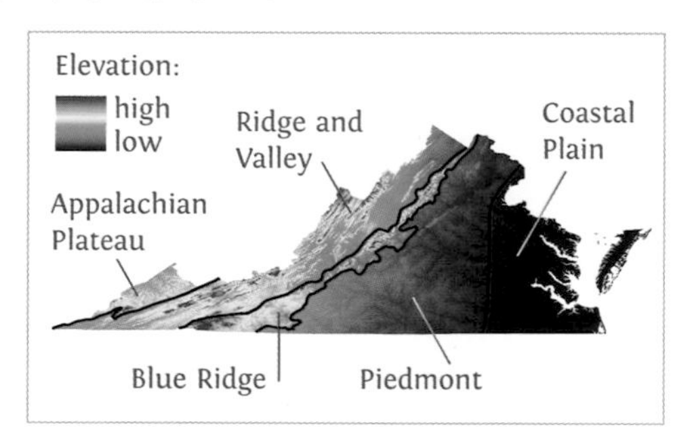

Physiographic Provinces

Rock type and hardness shape the physical and chemical characteristics of freshwater habitats, which in turn influence the distribution and evolution of fishes. Virginia has five landforms, termed physiographic provinces (see above), with distinct geologic histories. The following characteristics, habitats, and representative fishes of each physiographic province are listed in order of west to east.

Appalachian Plateau:

- Geology: Sedimentary sandstones and interbedded shale.
- Topography: Sharp, vertical mountains rising up to 2,500 feet above sea level and steep, narrow valleys.
- Stream Types: Stair-steps of plunge pools and rapids that run clear, cold, and high in oxygen. Water is typically soft due to the lack of carbonate rocks. Stream substrate is dominated by boulder, cobble, and bedrock.
- Fish Diversity: Low to moderate.
- Representative Fish Species: Blacknose Dace and Sculpin.

Ridge and Valley:

- Geology: Erosion-resistant sedimentary sandstones cap the ridges. Shales and carbonate rocks of limestone and dolomite dominate the valleys.
- Topography: Wide sweeping valleys between steep, parallel mountain ridges reaching over 4,700 feet above sea level.
- Stream Types: Fast-flowing streams on the ridges have similar physio-chemical characteristics to those in the Appalachian Plateau. Being less constricted by mountains and with lower gradient, valley streams begin to meander laterally to create distinct pool, riffle, and run habitats. Smaller stream substrates of pebble and gravel become prevalent. Water is hard due to the dissolving of carbonate rocks.
- Fish Diversity: While fish species in the ridges are similar to those in the Appalachian Plateau, valley fish diversity is high due to increased habitat diversity, warmer temperatures, and higher productivity.
- Representative Fish Species: Smallmouth Bass, redhorse suckers, and various darters and minnows.

Blue Ridge:

- Geology: Highly erosion-resistant granite, quartz, and greenstones.
- Topography: The Blue Ridge Mountains occur in this province, including Virginia's highest peak, Mount Rogers, at 5,729 feet above sea level. Narrow mountain peaks dominate the northern portion that broadens into a wide plateau southward.
- Stream Types: On the southern plateau, streams behave similarly to those in the valley portion of the Ridge and Valley region, but water is soft due to the lack of carbonate rocks and cold due to the high elevation. Streams that flow off the plateau or mountains behave similarly to those in the Appalachian Plateau.
- Fish Diversity: Low to moderate.
- Representative Fish Species: Brook Trout, Rainbow Trout, and Brown Trout are abundant in Blue Ridge streams. The Kanawha Darter is unique to this province.

Piedmont:

- Geology: Erosion-resistant igneous and other metamorphic rocks.
- Topography: The Piedmont is the largest province in Virginia and drains entirely into the Atlantic Ocean. It contains gently rolling mountains with elevations that vary east to west 200 to 1,000 feet above sea level.
- Stream Types: In the western Piedmont, streams have moderate gradients, forming distinct riffles, runs, and pools. Moving eastward, streams begin to slow and meander, forming short riffles and long runs and pools. In the easternmost portions of the Piedmont, streams become slow and swampy—characteristics of Coastal Plain streams. Because the main substrate is sand and silt, woody debris (i.e., fallen trees, branches, etc.) provides important habitat for fishes in this area.
- Fish Diversity: Moderate to high.
- Representative Fish Species: Stripeback Darter and Roanoke Bass.

Coastal Plain:

- Geology: Primarily sand and clay.
- Topography: Elevation is the highest at 200 feet and averages less than 100 feet above sea level.
- Stream Types: Low gradient produces slow, meandering streams flowing through swamps and flooded forests. With a few exceptions, water in this region is acidic, dark, and low in oxygen. Stream bottoms are comprised of mud, sand, and decaying plant matter. Submerged aquatic vegetation can become abundant in open areas of beaver ponds and millponds.
- Fish Diversity: Low to moderate.
- Representative Fish Species: Swampfish, Bluespotted Sunfish, and Pirate Perch.

Drainages

Overlaid onto Virginia's five physiographic provinces are two major basins: the Mississippi River and Atlantic Slope. Water draining to the Mississippi River basin eventually enters the Gulf of Mexico. Atlantic Slope rivers empty into the Chesapeake Bay or the Atlantic Ocean. These two basins are divided into ten smaller drainages. The Roanoke, James, Chowan, Potomac, Rappahannock, York, and Pee Dee rivers are in the Atlantic Slope and cover approximately three-quarters of the state. The Big Sandy, Tennessee, and New rivers are in the Mississippi River basin and cover the remaining land area. General information (size, major river systems, impoundments, and species richness) for Virginia's drainages is provided in Table 1.

Among Virginia's drainages, native fish diversity is highest in the Tennessee and lowest in the Pee Dee (Table 1). Many species are widespread and occur in multiple drainages. Other species, due to their evolutionary history, occur in a single drainage. These species are termed endemics and by drainage, the Tennessee has the most with 16, followed by the New (8), Roanoke (6), James (3), and Potomac (1). Examples of endemics include Blotchside Logperch (Tennessee), Orangefin Madtom (Roanoke), and Longfin Darter (James). In all drainages, the actual number of species has been artificially increased, intentionally or accidentally, via human introductions. The most extreme example of introductions would be the New drainage, which has an estimated 40 native and 46 introduced species.

The ten drainages of Virginia:

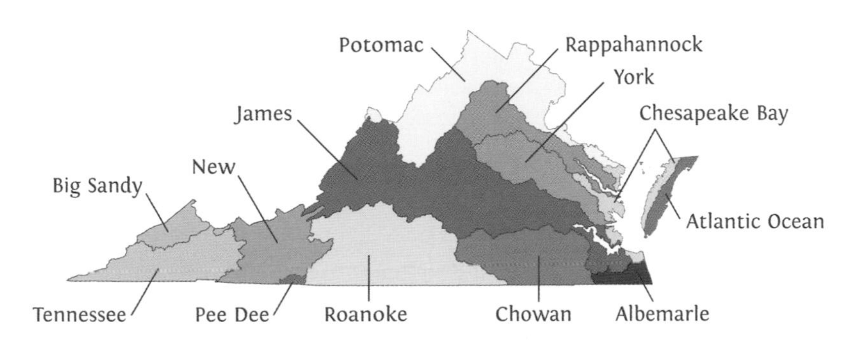

Table 1: Drainage information for the ten major drainages in Virginia

Drainage	Total Area (mi²)	% in VA	Major Systems	Major Impoundments	Native Species
Tennessee	40,876	7.6	Powell, Clinch, Holston	South Fork Holston Lake	97
Potomac	14,700	38.8	Shenandoah	None	64
New	12,223	25.1	Little River, Walker Creek	Claytor Lake	40
James	10,432	99.3	Jackson, Rivanna, Appomattox	Lake Moomaw	79
Roanoke (Staunton)	9,680	64.9	Pigg, Banister, Big Otter Creek	Smith Mountain Lake, Kerr Reservoir	65
Pee Dee	7,221	1.6	Ararat	None	22
Chowan	4,800	84.6	Nottoway, Blackwater, Meherrin	None	71
Big Sandy	4,140	63.1	Russell Fork, Levisa Fork	Flannagan Reservoir	41
Rappahannock	2,715	100	Rapidan	None	60
York	2,661	100	Mattaponi	Lake Anna	57

Rivers and Streams

From a raging mountain stream to the deepest, darkest swamp, fishes occupy most of Virginia's freshwater habitats. By far the most abundant and widespread are the flowing systems of streams and rivers. Over 27,000 miles of streams and rivers flow through the Commonwealth. At 348 miles long, the James is the longest river within Virginia, where it crosses four physiographic provinces and its watershed encompasses 25% of Virginia's total land area.

Rivers and streams can be classified as warm, cool, and cold systems depending on the maximum average summer water temperatures. Temperature is related to latitude, elevation, land use, and groundwater influences. Certain species, such as Brook Trout, are exclusive to cold-water systems with temperatures less than 70°F. A warm-water system contains species that can tolerate temperatures over 81°F, including bass, sunfish, and catfish. Cool water has the fewest species and intermediate temperatures. Walleye and certain darter species are good examples of cool-water fishes.

Rivers and streams contain three basic habitat units—riffles, pools, and runs. Riffles are caused by a sudden drop in elevation resulting in a shallow, fast-flowing reach where water is turbulent. In larger rivers, riffles can become rapids, which are much deeper and faster than riffles. Pools are deep, slow-moving sections of flat water. Runs have characteristics of pools and riffles with velocities and depths intermediate between both. Whereas some fish species spend their entire lives in one habitat unit, others will use multiple units, depending on the season and their life history characteristics.

Mountain stream riffle

Piedmont stream run

Ridge and Valley pool

Coastal Plain millpond

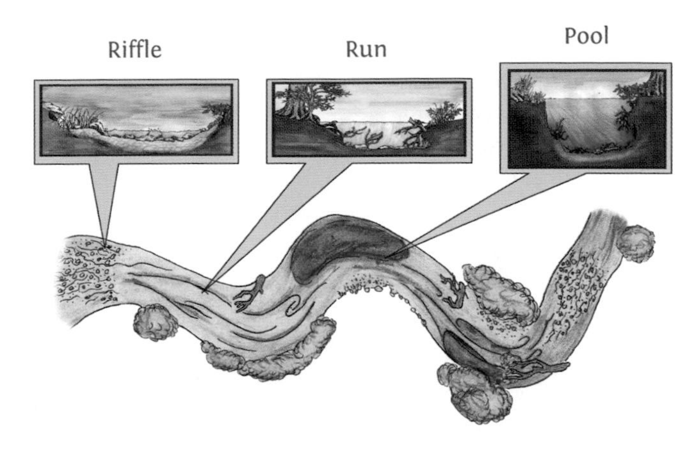

Lakes, Ponds, and Reservoirs

Other freshwater habitats in Virginia are lakes, ponds, and reservoirs. Only two natural lakes occur in Virginia: Lake Drummond, in Suffolk and Chesapeake cities, and Mountain Lake, in Giles County. Other lakes and most ponds are impounded segments of rivers and streams. The largest of these is John H. Kerr Reservoir, an impounded portion of the Staunton River in Mecklenburg County. Other major impoundments by drainage are listed in Table 1. Of the natural ponds that do exist, most are created by beavers or are natural sinkholes. Whether created by beavers or humans, lake and pond habitats will have different fish species.

Ponds and lakes lack constantly moving water like rivers and streams. In large lakes, the water will stratify during the summer into distinct (upper, middle, and lower) layers depending on the time and temperature. The upper layer (or epilimnion) has the most sunlight penetration, which causes it to become warm, high in oxygen, and extremely productive. Many fish species, such as Threadfin Shad, Largemouth Bass, and crappie, will inhabit this layer. In contrast, the lower (or hypolimnion) is cold, dark, and low in productivity. In the deepest portions of the lake, oxygen can become depleted due to the decomposition of organic material. If oxygen levels are sufficient, fishes such as trout will inhabit this layer to avoid warm temperatures near the surface. Other fish species prefer the narrow middle (or thermocline), which offers intermediate temperature, light, oxygen, and productivity levels compared to the other layers. Striped Bass prefer this layer during the summer because of the cooler temperature and necessary oxygen levels. As the summer progresses, this layer will decrease in thickness, causing Striped Bass to become increasingly stressed. Turnover occurs during fall and spring, causing the mixing of all layers as a result of wind action and temperature changes.

Ecological Importance

Natural systems rely on complex and dependent relationships among species to operate effectively. One such relationship is the food web or chain, in which nutrients and energy are transferred from one feeding (trophic) group to another. Fishes are an integral component of this relationship, providing an energy link between their food and the aquatic and terrestrial species that feed on them.

Example of an aquatic food web:

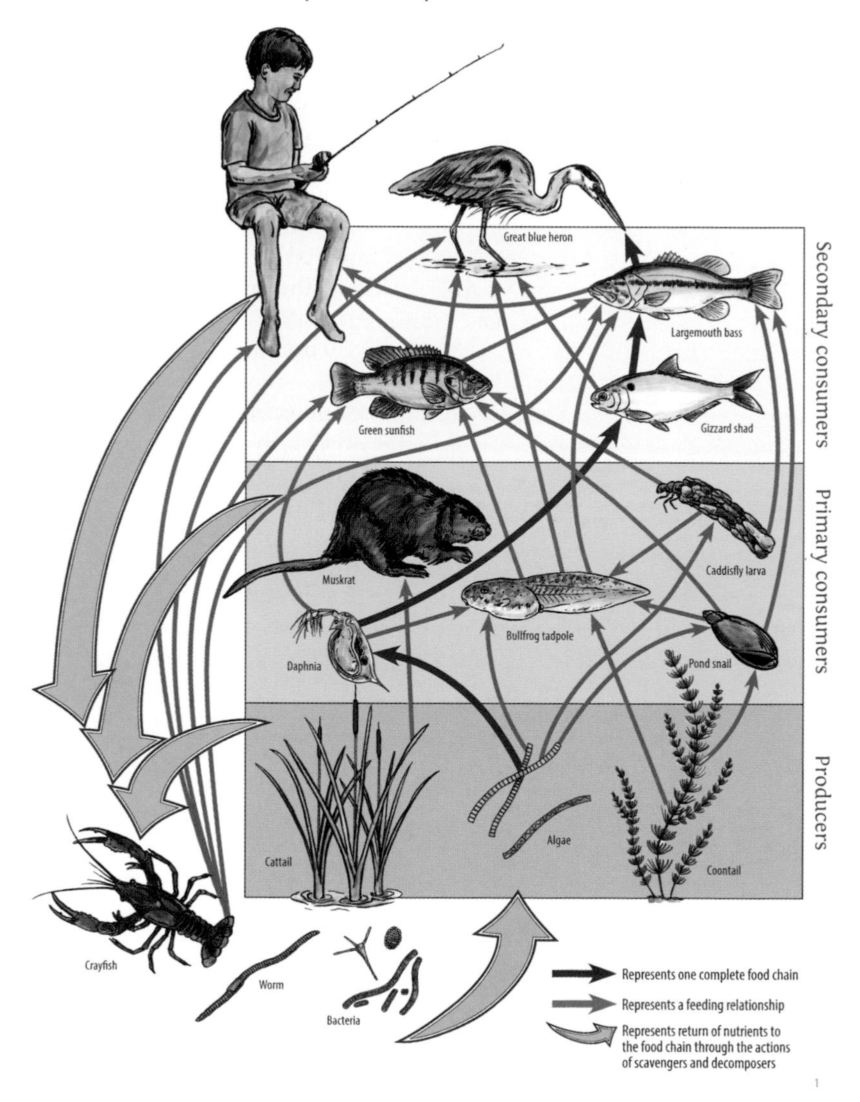

In most aquatic systems, the foundation is microscopic algae. The Central Stoneroller, a stream minnow, specializes in eating algae attached to rocks and earned its name by turning over stones while feeding. In lakes and large rivers, Threadfin Shad perform a similar function and feed on free-floating algae and microscopic animals collectively called plankton. Virginia's most extreme plankton-eater is the Paddlefish, a primitive 5-foot-long species that feeds like a baleen whale by moving through the water column with its enormous mouth open.

Many fish species occupy higher levels of the food chain and feed on larger organisms—aquatic and terrestrial insects, mollusks, and crustaceans. As the fly-fishing angler knows, Brook and Rainbow trout feed heavily on aquatic insects such as mayflies, stoneflies, and caddisflies that in turn feed on detritus and algae. The Greenside Darter feeds on freshwater snails and other aquatic invertebrates.

Atop the food chain are fish species that feed mostly on large animals and other fishes. Equipped with a large toothy mouth, the Muskellunge feeds on suckers, minnows, and sunfish. In order to maintain a top predator, the base of the food web must be healthy and stable. One example would be the Smallmouth Bass that feeds on both aquatic insects and small fishes. It is estimated that a bass would have to consume over 200,000,000 aquatic insects or 20,000 minnows to gain 1 pound.

Beyond serving as predator and prey, fishes have other interdependent relationships with their environments. During the spawning season, the River Chub, Fallfish, and other minnows build elevated mounds made of stones in rivers and streams. Reaching up to 1 foot high and 3 feet in diameter, the mound is a natural incubator, allowing water to flow between the stones, providing oxygen and gently turning the eggs inside. Not only does the mound builder use it to spawn, but up to 24 species of Virginia's other minnows (nest associates) are known to use these mounds, either exclusively or intermittently. In this manner, mound builders are considered keystone species, because they can dramatically change the species composition in streams.

Fishes are uniquely interconnected with other animal groups as well. One example is their relationship with highly imperiled freshwater mussels. To complete their life cycle, most freshwater mussels need to attach their larvae (glochidia) to a freshwater fish. While some mussels can use multiple fish species, others are specific to just one or two. Once attached, the larva non-lethally parasitizes the fish and develops into juvenile mussel. After several weeks, the juvenile then detaches from the fish to begin its free-living life on the bottom of a river or lake. Because mussels are sedentary, the fish serves as a taxi to disperse the young mussels to other areas. Although this relationship seems one sided, fishes do benefit. Mussels feed by filtering suspended particles, thereby enhancing water quality. Furthermore, mussels are fed upon by multiple fish species, including Freshwater Drum and redhorse suckers. Even the empty mussel shells provide shelter and egg-laying sites for small catfishes like the Mountain Madtom.

Freshwater mussels require a host fish to complete their life cycle. The Wavy-rayed Lamp mussel, *Lampsilis fasciola* (above), uses a darter-like lure to attract its fish host.

Economic Importance

Freshwater fishes provide important contributions to Virginia's economy. One of the most popular activities related to fishes is recreational fishing. The US Fish and Wildlife Service estimated in 2011 that 8% of Virginia residents fished a total of 9,367 days and spent $2.6 billion. Besides the revenue generated from the purchase of boats and gear, fishing provides an economic multiplier related to the cost of travel, meals, and other amenities. Many local jobs from the guide service operations, tackle shops, and boat rentals are supported by fishing opportunities. One notable benefit in the sale of boats and fishing equipment is that a small taxable portion goes directly to the state fish and wildlife agency to better manage aquatic resources. In Virginia, the Department of Game and Inland Fisheries (VDGIF) manages freshwater resources in cooperation with the US Fish and Wildlife Service, the US Forest Service, and other public and private entities.

An angler participates in a Virginia youth fishing event.

FISH WATCHING, PHOTOGRAPHING, AND KEEPING

Fish Watching

From Virginia's mountains to valleys, you can stand on a shore and make a hobby of watching fishes. Whether watching trout sip insects from the water's surface or being dazzled by spectacular spawning minnows on the nest of a chub, the casual observer can be drawn into a mysterious underwater world.

Saffron and Tennessee shiners on a River Chub spawning mound

Students snorkeling to observe Candy Darter in chilly mountain water

Simply put on a pair of polarized sunglasses (to cut through the surface glare) and stroll or float along a creek or pond. Or use a viewing bucket, a 5-gallon bucket with the bottom replaced with a pane of plexiglass, to peek below the water surface without glare. Other amazing forms of aquatic life can be seen, such as turtles, salamanders, crayfish, aquatic insects, mussels, and even freshwater sponges.

Many of Virginia's waters have clear, warm water suitable for viewing with a mask and snorkel. Instead of being scared, many fishes become bold and curious about a large creature immersed in the water rather than looming over them like a hungry heron.

Visit a dive shop and try out masks and snorkels to find those that fit best. A wetsuit is also a good investment, allowing you to spend long periods of time immersed in cooler water. Wading boots are used instead of flippers because time is spent crawling in shallow, flowing water rather than swimming.

Snorkeling has inherent dangers. Some rivers have complicated and unpredictable flow patterns, hidden obstructions, and hazardous trash. Avoid murky water, be careful where you place your hands, wear gloves, avoid snags, and never snorkel alone. If you are a beginner, partner with a proficient snorkeler, talk to the staff at your local dive shop, and/or join a "river snorkeling" community. River snorkeling has gained popularity for blending fish observation with "extreme sport."

The best time to see interesting fish behaviors is during spawning season, when you may encounter round, light depressions in the bottom, indicating nesting sunfish, which often build nests in dense colonies. Nesting male sunfish make territorial displays as they compete for nest sites and protect their own nests. They also court females in established territories and may habituate to your presence.

Probably the most incredible spawning spectacle in our waters is the minnow spawning aggregation. When large, nest-building members of the minnow family like chubs and stonerollers start shifting gravel in order to construct clean spawning sites, a frenzy ensues. Nest-building minnows attract groups of smaller minnows, known as "nest associates," with extremely vibrant coloration. Some nests may have half a dozen or more species crowded around, spawning all at once and displaying coloration one might expect to see only on a tropical reef. Once focused on spawning, these fishes lose almost all fear of predators as the urge to spawn dominates their behaviors.

Saffron and Tennessee shiners as seen below the surface

A photographer using underwater camera housing and gear

Photography

While people have been diving around coral reefs with specialized camera gear for decades, more recently fresh waters are also explored and recorded in video and still photos. For a relatively low cost, you can find a rugged compact camera that is waterproof, is easy to use, and has great video resolution.

An important consideration for obtaining good photos is water clarity. Like snorkeling in general, the best time for underwater photography is during periods of low flow when rain has abated for at least several days. Finding sites with protected watersheds and quality habitat becomes especially important, as even in clear water, river bottoms can be brown and silty. Sunfish and darters are naturally curious and relatively easy to photograph. A skittish subject requires time to let it get comfortable in your presence and makes underwater photography fun and rewarding. Persistence and slow, deliberate movements can turn even the most nervous fishes into great photo subjects.

To admire the form and color of a fish, or to document species seen or caught, another method is to use a small aquarium or photo tank to isolate the fish. A variety of setups can be used, including plastic animal tanks or small glass aquariums. The more advanced and user-friendly photo tanks are often narrow enough to just fit a fish inside, with an internal V shape that wedges the fish inside and positions it upright.

The use of a photo box such as this not only keeps the fish safely in the water to reduce stress but allows for longer time spent viewing the fish. Submerged in water, many more details of the fish are readily seen. Gently tapping the tank can coax the fish to spread its fins. It may be desirable to bring clean, dechlorinated water to use for the best photographs. In a small container of water without flow, fishes may suffer from oxygen depletion and temperatures may quickly increase, especially in the sun. Be conscious of these factors, and in general, watch the fish closely to be sure it is still energetic and vigorous, and release it back where it came from or into an aerated bucket before it shows signs of stress or deterioration.

A photo tank used to view the spectacular Bluespotted Sunfish

Researchers kick the stream bottom and use a seine to capture fishes

Catching Fishes

Anyone with an in-depth interest in fishes will likely want to see a larger variety of species than what is typically caught with rod and reel. It's becoming more common for people to keep a "life list" of fish species. The very nature of listing these encounters can encourage friendly competition among enthusiasts and becomes an avocation that promotes continued efforts to find as many different fishes as possible. Others may be interested in monitoring fish diversity in a local creek as a means to assess the health of a waterway. Whatever the reason, there are many other ways to catch fishes besides traditional angling.

Before attempting to catch or keep any fish, purchase the required fishing license and review and abide by any applicable regulations governing the capture, possession, and transport of wild fishes. In Virginia, up to 20 nongame fishes may be taken.

Once removed from the water, they may not be released back to the wild or sold under any circumstances. Take care to not collect fishes in any prohibited areas, and do not tkae any protected species. If you cannot identify the fish, you should not consider removing it from the wild. Be sure to check with the Virginia Department of Game and Inland Fisheries for any changes or updates to these regulations.

"Micro-fishing" is the practice of catching fishes of small size and variety. Using special tiny hooks, tiny baits, usually a fixed length of line, and a very light rod, this activity has exploded in popularity. Micro-fishers often target new species to increase the challenge and sense of accomplishment. Photos are taken of fishes in hand or in a photo tank as proof of the catch, before the fishes are released unharmed (see p. 25).

Fishes may also be caught by means of nets. Seine nets are rectangular nets stretched between two poles and are perhaps the most common, efficient means of catching fishes (see p. 25). In Virginia, seine nets may legally be as large as 4 feet tall by 10 feet across and can be used in various ways, from actively "hauling" the seine into pockets of water or by placing it in the current perpendicular to water flow and kicking and disturbing the rocks. This kick seining will coax darters and other benthic fishes into the net. A dipnet or large aquarium net may also be helpful when trying to capture fishes in swampy areas, where they are often hiding in vegetation.

Dipnets of various sizes can be used like small, personal seines when placed downstream if the stream is disturbed. Cast nets are more often used by anglers to collect larger pelagic fishes. When placed appropriately, minnow traps can also be productive. Legally, these must have openings no larger than 1 inch in diameter, to avoid accidental capture of larger fishes. Baited with bread or canned cat food, placed in a slow, deep current, after a period of time a trap may be full of fishes. If left for more than an hour or two, a minnow trap may also attract predators such as crayfish, snakes, or curious mammals looking to capitalize on a concentrated group of fish, especially after dark. This interest in a much wider array of fishes has the secondary effect of helping participants learn more about fishes they might not otherwise know existed.

It is best to transfer a catch to a home aquarium in a cooler with plenty of the water it was collected in and supplied with oxygen and circulation by means of a battery-powered air pump. Add commercial aquarium salt to the water immediately after collecting fish. This helps the fish maintain osmotic balance and also encourages the production of more mucus, which protects the fish from pathogens.

Transporting is least stressful in times of cool weather before spawning or in autumn. Wintertime collection requires the fishes have a very slow (at least a day or two) acclimation to household temperatures. Avoid collecting adults during spawning, as these fishes are already stressed and unlikely to survive collection and should be allowed to complete spawning without interruption for the sake of the next generation.

Fish Keeping

Most of our native fish species, especially the darters and shiners, are quite colorful and stay small enough to be easily maintained in a home aquarium. While the aquarium hobby is a fun and rewarding pursuit, it is also quite complex. Books such as *American Aquarium Fishes* are helpful references. Here we provide the basics of how native fish keeping differs from traditional tropical aquarium keeping.

The aquarium should be designed to approximate the habitat for a target fish. Water chemistry, amount of water movement, bottom composition, and amount of cover or hiding places are essential to health and success. Aquatic plants add not only beauty to the aquarium but also stability, as healthy aquatic plants can help sequester excess nutrients and provide enhanced oxygenation in the aquarium.

A Greenside Darter perches naturally in an ideal aquarium habitat

Sunfishes are beautiful aquarium fishes but can be aggressive

Once an aquarium is set up and is properly running (a process similar to that of a tropical aquarium, without a heater), the fun really begins. Minnows or darters do well, provided that the water is well oxygenated. With a mixture of small minnows and shiners there is less chance of predation in the aquarium, unless fishes of different sizes are mixed. In an aquarium setting, shiners will almost always accept commercial flake fish food as a staple diet, while darters typically require either live or frozen foods such as brine shrimp, *Mysis* shrimp, or bloodworms.

Sunfishes are another popular choice for native fish enthusiasts. Most species are brightly colored, with some rivaling the most attractive tropical fish species. Small sunfishes in the genus *Enneacanthus* can also be good aquarium fishes, require less space, and have unusually docile temperaments. While sunfishes will coexist with tankmates that are too large to eat, they are more predatory and territorial, especially with other sunfishes. Sunfishes often seem to have individual personalities and hierarchies. A stable aquarium with a few fishes that get along may be thrown into temporary chaos with the addition of a new tankmate, as pecking order must be reestablished.

Many fishes other than the minnows, darters, and sunfishes have the potential to be interesting aquarium fishes. Be sure to adapt the aquarium and feeding to match their requirements. Native fish aquariums often feature a single fish species, for which the habitat is replicated in as much detail as possible. Single-species aquariums offer the best chance to discover interesting behaviors or encourage breeding. While difficult for most home aquarists, breeding native fish species can be fun. However, the offspring may not be legally sold or returned to the wild. Our native species are adapted to seasonality and will require temperature shifts and photo-period simulation to breed.

While native fish keeping is a small niche of the aquarium hobby industry, there are resources geared toward the keeping of native fish species. The North American Native Fishes Association (NANFA) is well known and perhaps the best group around for anyone interested in keeping native fishes in captivity and for promoting all of the activities covered in this chapter (and much more!). There is a wealth of useful information in their online forums, in their quarterly publication, *American Currents*, and in the expertise of their members.

The Virginia Master Naturalist program offers opportunities to learn about local fish species as well as many other facets of the natural world. If you wish to get involved in assisting with fishery science or conservation in your area, volunteering can be a fantastic way to make a conservation impact while also having positive experiences with fishes in the wild. The VDGIF, Trout Unlimited, and other conservation organizations generally appreciate enthusiastic volunteers to assist with field work or habitat restoration. As an enthusiast of native fishes, you may also wish to volunteer for a "BioBlitz," a celebration of biodiversity whereby various groups of plants and animals are surveyed in a particular area in order to create an inventory of species.

The pursuit and enjoyment of our diverse fish fauna can be a fun and educational endeavor that can help connect us with nature and science in numerous ways. As Steven A. Forbes said in 1887 in *The Lake as a Microcosm*, "If one wishes to become acquainted with the black bass, for example, he will learn but little if he limits himself to that species. He must evidently study also the species upon which it depends for its existence, and the various conditions upon which these depend. He must likewise study the species with which it comes in competition, and the entire system of conditions affecting their prosperity . . . "

MANAGEMENT and CONSERVATION

Introduction

How does a scientist "manage" fishes? Is a harvest sustainable? How do we make sure enough fishes remain for future generations? How will we deal with climate change and human population growth? These and other questions are addressed through scientific inquiry that determines the health and balance of Virginia's freshwater fishes.

Background

The early years of freshwater fisheries management emphasized fish culture to restore fish populations. The Wytheville Hatchery, built by the Virginia Fish Commission in 1879, was the first hatchery in Virginia to raise cold-water species such as trouts and salmons. Pacific salmon were released into James River and Roanoke River watersheds as early as 1875. Although well intentioned, raising a west coast migratory fish for Virginia was unsuccessful and soon abandoned. Propagation of European Brown Trout soon followed.

Populations

Fisheries managers inventory fish populations to ensure quality fishing. Fishery biologists identify species, count numbers, and examine where they move and what they eat. A wise fisheries biologist once said, "Managing fisheries is hard: it's like managing a forest in which the trees are invisible and keep moving around."

Fisheries biologists use electrofishing to sample a stream

Radio tagging provides opportunities to learn about movement patterns

Sampling

Fisheries biologists often target certain species, selecting appropriate sampling gear to obtain a sample reflective of the population, and apply mathematical models to predict future trends. Nets are usually involved when sampling fish. A type of net used in freshwater sampling is the seine. Seines, trap nets, gill nets, trawls, and electrofishing are favored methods for sampling freshwater fishes.

Tagging

Marking or tagging fishes is used extensively. Tags can be attached or inserted into fishes to document growth and movement, or they can be used to batch-mark cohorts of fishes to follow them through life. Marking fishes can be as simple as using a caudal-fin clip, external anchor tags, or surgically implanted radio tags. Tags permit anglers to report size and harvest, and biologists measure exploitation, growth, or movement.

Recruitment, Growth, and Mortality

Fisheries biologists must have a good grasp of population recruitment, growth, and mortality to develop regulations and management strategies for quality fishing. Evaluating recruitment success or failure is achieved by determining the number of young fishes entering the fishery on an annual basis. Because most freshwater fishes spawn once a year, understanding year-class strength of a species is an essential component of fisheries management. Fishes grow throughout life, but the young grow rapidly, and a slowdown is inevitable. Scientists use structures, such as scales, to estimate age. When observed under a microscope, a scale looks like tree rings, with rings radiating from the center, or "focus." In winter, these radii bunch up to form an "annulus." The biologist who has recorded a corresponding length of the fish can then match size to age. Otoliths (ear stones) are now typically used to determine ages for Virginia freshwater fishes because they are more reliable than scales. These clues to age structure, coupled with creel surveys, permit the fisheries biologist to estimate total mortality and fishing mortality and to gauge the appropriateness of current regulations.

Diet

Most anglers want to know, "What are they bitin' on?" As with any wild animal, it is important to know feeding habits, as they relate to ecological impact or the growth of targeted sport fish. Predator-prey interactions, particularly in reservoirs, have been studied extensively to understand energy flow and how management can address slow growth rates or low sport fish weights. Sometimes resource agencies stock fishes to supplement existing fisheries or establish new fisheries. Stocking an apex predator into a river system to provide trophy fish angling requires an abundant supply of prey fish.

Angler Surveys

Data on sport fishers' attitudes and effort are obtained through angler surveys that directly engage anglers as they are fishing. Questions vary between locations, and the answers yield valuable information about catch rates, fishing effort, and harvest (removal) of targeted sport fish. Important demographic and economic data are also gathered to inform regulatory decisions regarding fishing satisfaction.

Health

Scientists react to fish kills by conducting investigations exploring results of human-caused pollution events or a crash in dissolved oxygen levels. Some fish kills are quickly solved cases, and the damaged stream or river quickly rebounds. The more challenging fish health issues deal with chronic problems, such as bacterial or viral infections, contaminant burden, heavy parasite infestations, tissue rot, blindness, and melanosis. The 1972 Federal Clean Water Act regulates pollutants that originate from a pipe. Non–point source pollution occurs from runoff from impervious surfaces, farm fields, lawns, and mines. Forests often introduce excessive nutrients, sediment, and chemicals to watersheds. These multiple stressors directly or indirectly reduce fish health. Natural filters, such as healthy riparian zones and wetlands, may trap toxicants before they reach surface waters.

A stream reach prior to restoration

The same stream reach after restoration

Genetics

Advances in genetic research have given biologists tools to identify subpopulations of targeted fishes with just a small tissue sample. Genetic variation provides clues about how and when fishes are changing and adapting over time. Additionally, hatchery managers can genetically modify fish eggs and create sterile offspring that grow better in hatcheries and do not reproduce in the wild. Hatchery-produced hybrid fishes are commonly stocked to enhance sport fishing.

Habitat

Healthy physical habitat is essential for fishes to survive and thrive. Habitats may be improved by adding attraction structures. Anglers fishing these areas will have a greater opportunity to catch game fishes. Removal of low-head dams that have outlived their usefulness is another way to restore river function, provide public safety, and promote fish passage. Other types of fish passage barriers are culverts. Bad culvert design can isolate fish populations and reduce population growth.

Threatened and Endangered Species

The last documented Virginia sighting of the Harelip Sucker (*Moxostoma lacerum*) in the North Fork Holston River was made in 1888. Meanwhile, the last collections of Trout-perch (*Percopsis omiscomaycus*) in the Potomac River drainage were in 1948.

The Harelip Sucker is now extinct, but Trout-perch are still present in northern states and Canada. To prevent further extinctions and extirpations, Virginia currently has identified 23 fish species as endangered or threatened. Listed species cannot be collected or possessed, nor can their habitat be altered or destroyed. Management of listed fish species is the responsibility of the VDGIF and the US Fish and Wildlife Service.

Recovery efforts for endangered species include species surveys, population monitoring, life history determinations, habitat associations, habitat restoration, propagation, and reintroduction. To accomplish recovery, state and federal management agencies work cooperatively with other governmental, nonprofit, and academic organizations and institutions.

Hatcheries and Stocking

Five VDGIF state cold-water hatcheries stock close to 1.5 million trout for recreation on an annual basis. Raising and stocking non-indigenous species such as Brown Trout and Rainbow Trout has led to their naturalization in some waters where the native Brook Trout thrive. Interspecific competition and hybridization with Brook Trout are two reasons why fisheries biologists are cautious when utilizing these species in state waters. Four VDGIF state warm-water hatcheries stock close to two million Striped Bass, Walleye, and Channel Catfish for recreation on an annual basis. Stocking artificially raised fishes in artificially constructed environments, such as reservoirs, has proven to be successful, and some cannot be sustained without annual stocking of hatchery-reared fingerlings. Many of Virginia's public fishing lakes do not have sufficient spawning habitat for Channel Catfish, so they are supplied with annual allotments of advanced fingerlings in order to sustain high-quality fisheries. VDGIF raises approximately a million New River Strain Walleye that replenish stocks in the New River and across the state. Additionally and in partnerships with other governmental, nonprofit, and academic institutions, VDGIF has been responsible for the reintroduction of the endangered Yellowfin Madtom.

What You Can Do

1) Know federal and state agencies and "Friends of" organizations that focus on the environment, how they work, and how to contact them about problems.

2) Participate in freshwater fisheries conservation by knowing your native fishes, practice catch-and-release, and do not move any fish from one habitat to another.

4) Before stocking any freshwater plant or animal in a watershed, discuss your plans with VDGIF district biologists.

3) Report fish kills, unauthorized channelization, or changes in water quality to the state Department of Environmental Quality.

4) Help take care of or restore riparian habitat on land holdings essential for clean water and wildlife connectivity.

5) Dispose of waste properly and leave waterways cleaner than when you arrived.

FISH ANATOMY

One needs to know some basic terms in order to identify the fishes of Virginia. Whether you are streamside or lakeside, you can learn a lot by observing fishes and their anatomy. Each specimen tells a story about how it lives, feeds, and survives, even before you can name the specimen. Distinguishing characteristics used in this field guide include the body shape and proportions, location of fins, colors, and patterns. Other more tedious counts and measurements as well as internal anatomy may be used by specialists in the lab but are not covered here so you may spend more time outdoors!

Often size is important to aid in identification. Virginia fishes vary greatly in size, from small darters and minnows to the large Atlantic Sturgeon. Size is often correlated with longevity. Eastern Mosquitofish reach only 2 inches and live about 1 year, whereas the Atlantic Sturgeon reaches 14 feet and can live as long as 150 years. Beyond size alone, a close look at the body form and anatomy will reveal characteristics we use to identify the fish. Body shapes in profile may be fusiform (tapering), oval (disc-shaped), or eel-like. Body proportions in side view (i.e., depth relative to length) may be elongate, moderate, or deep, and body form in frontal view may be round, oval, compressed, or depressed. Some fishes, such as catfishes, may be depressed in the head region while the body is round in the middle and compressed near the tail.

Body shapes of fishes:

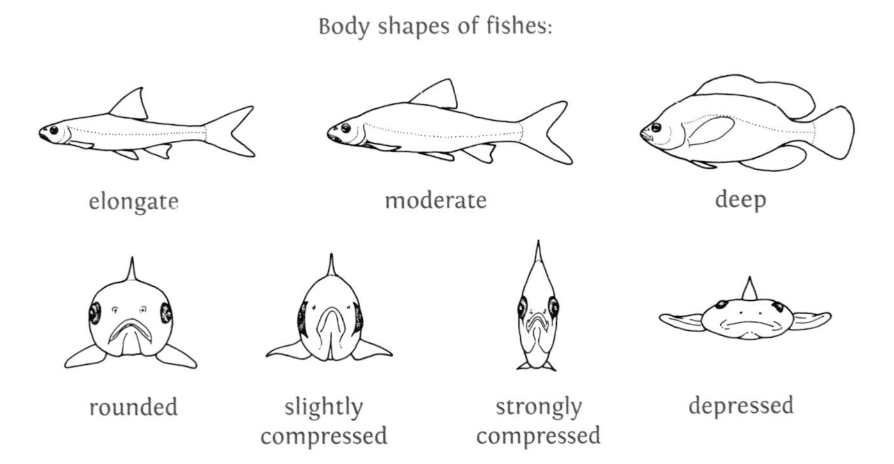

Descriptions of species in this field guide and the key to families use consistent terms to describe the anatomy of the fishes. Locations on the body of a fish are called anterior at the head end, posterior at the tail end, dorsal for the back, ventral for the belly, medial for the center line, and lateral for the sides.

Most fishes are counter-shaded, meaning they are dark above and gradually paler below. The color patterns are sometimes important for field identification of fishes. However, color can be a misleading indicator for identification because fishes may change color with age, sex, time of day, habitat, stress, or breeding condition. Minnows and darters in North American streams may appear drab much of the year, but in early spring the males change to brilliant colors to attract females for breeding.

Therefore, the following patterns are often more reliable traits than color. Stripes are patterns that run lengthwise from head to tail; these may be narrow or wide. Bands or bars are vertical patterns extending from the back to the belly. Sometimes bars are described by location on the body of the fish, such as humeral bar, located just behind the head, or the preorbital, suborbital, or postorbital bars around the eye. Saddles are spots or blotches that extend across the back and down onto the sides. Spots are distinct, with sharp edges, whereas blotches are diffuse. Blotches may be separate or conjoined in species of darters. Sometimes pigment patterns are shaped like crescents or stitches. Ocelli (singular: ocellus) are spots with an outer edge of contrasting color.

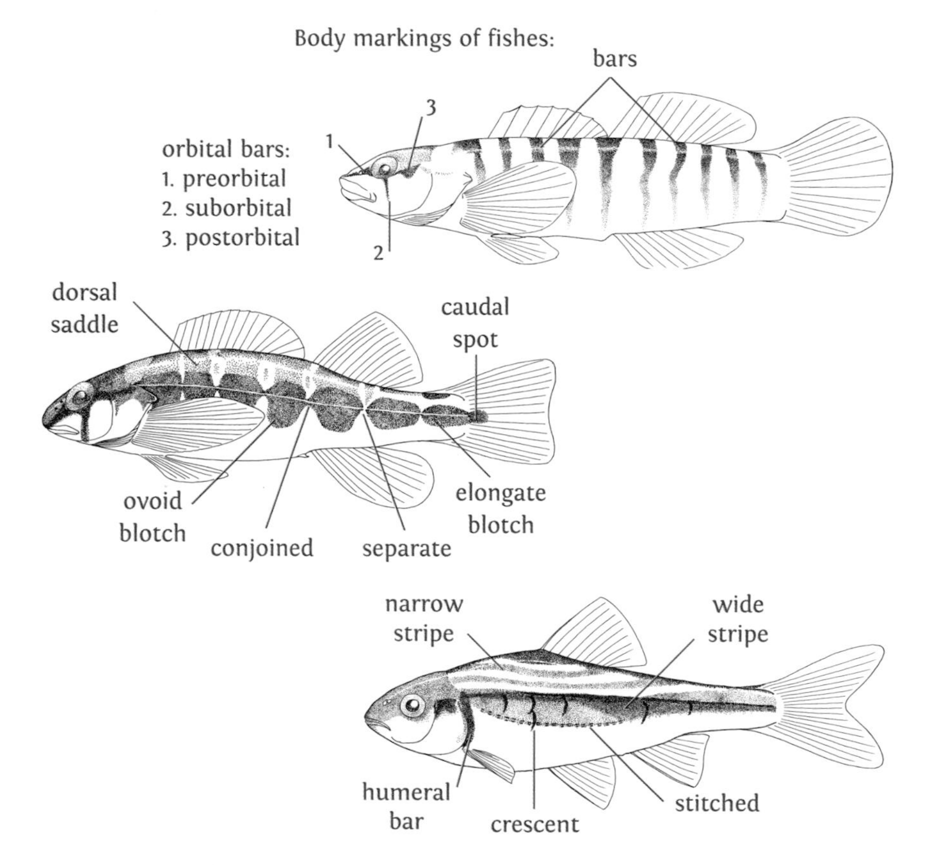

The head of a fish features eyes, nostrils (nares), gills, and mouth. Specific regions of the head are the isthmus, gular, chin, snout, cheek, and humeral. The isthmus is the narrow triangular area on the underside of the fish, anterior to the breast, that externally separates the gills. The gular region is the area between the lower jaws. The chin is the anterior part of the lower jaw. The snout is the area in front of the eyes and above the mouth. The nape is the region between the head and the dorsal fin. The cheek is the area below the eyes to the posterior end of the preopercle. The humeral region is just behind the operculum. The nostrils, situated ahead of the eyes, are olfactory organs, used only for smell. Gills are organs that permit fishes to obtain oxygen dissolved in the water and to diffuse wastes out of the body. While lampreys and sharks have multiple gill openings, in the bony jawed fishes, the operculum is the gill cover that protects the gills. This operculum is supported by the opercle and subopercle bones. The preopercle bone is just anterior to the opercle and has serrations along its margin in certain fishes, such as Yellow Perch. Fishes (excluding lampreys) have two jaws; the upper jaw consists of the premaxillary and maxillary bones and the lower jaw. In some ray-finned fishes, the premaxillary bones are capable of moving forward, making the mouth opening more tubular.

Parts of the head:

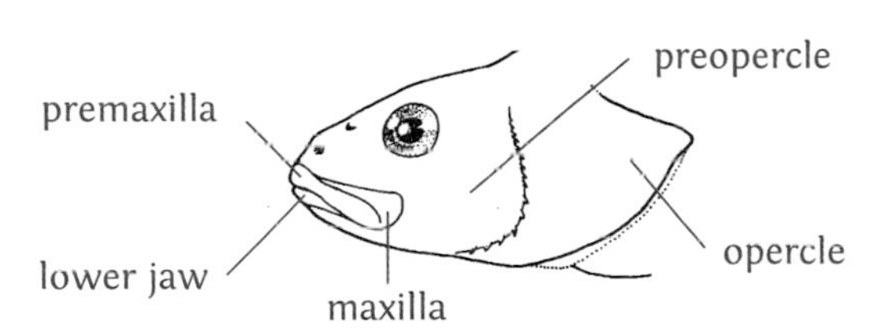

The position of the mouth varies among species and reflects feeding habits. Mouth positions are called terminal when the anteriormost points of the upper and lower jaws meet at the same point at the tip of the head. In the superior mouth position, the lower jaw projects beyond the upper, and the inferior mouth position is when the upper jaw projects past the lower jaw. Subterminal mouth position is when the upper jaw projects slightly beyond the lower jaw. Horizontal and oblique are terms used to describe the angles at which the mouths of fishes are oriented.

Mouth positions:

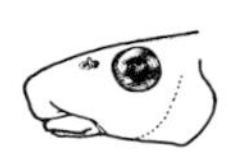

inferior, horizontal

subterminal, slightly oblique

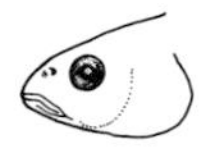

terminal, moderately oblique

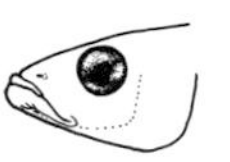

supraterminal

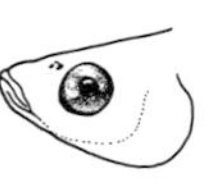

superior, stongly oblique

Fish fins are described as median (running down the midline of the fish) or paired (occurring in pairs). The median fins are the dorsal, caudal (tail end), and anal (just posterior to the anus) fins. Some families have an additional, small fleshy adipose fin on the midline of the back just behind the dorsal fin. The body tapers gradually to the caudal fin and this narrow region is called the caudal peduncle. Paired fins are pectoral and pelvic fins. Pectoral fins are attached behind the gill cover. However, pelvic fins can be in the abdominal (on the belly) or thoracic (below the pectoral fins) region, depending on which family the fish belongs to. The part of the fin joined to the body is called the base, whereas the forward limit of the fin base is called the origin and the posterior limit the insertion.

The number and location of fin rays and fin spines are diagnostic characteristics of fish species. All freshwater fishes of Virginia except the primitive lamprey are termed ray-finned fishes. These fishes have fins supported by long, thin, bony or protein-based rays connected by webs of skin. Rays are typically flexible and segmented, allowing for more precise maneuvering. Spines are present in catfishes, sculpins, basses, sunfishes, perches, and drums. Spines are hollow, unsegmented, and often sharp and occur in the anterior portion of pectoral, dorsal, pelvic, and anal fins of some fishes, called spiny-rayed fishes. Fishes may possess one or two dorsal fins that may be spiny (with spines) or rayed. The spiny and rayed portions of the dorsal fin may be connected, or there may be a deep notch.

Parts of the body and fins:

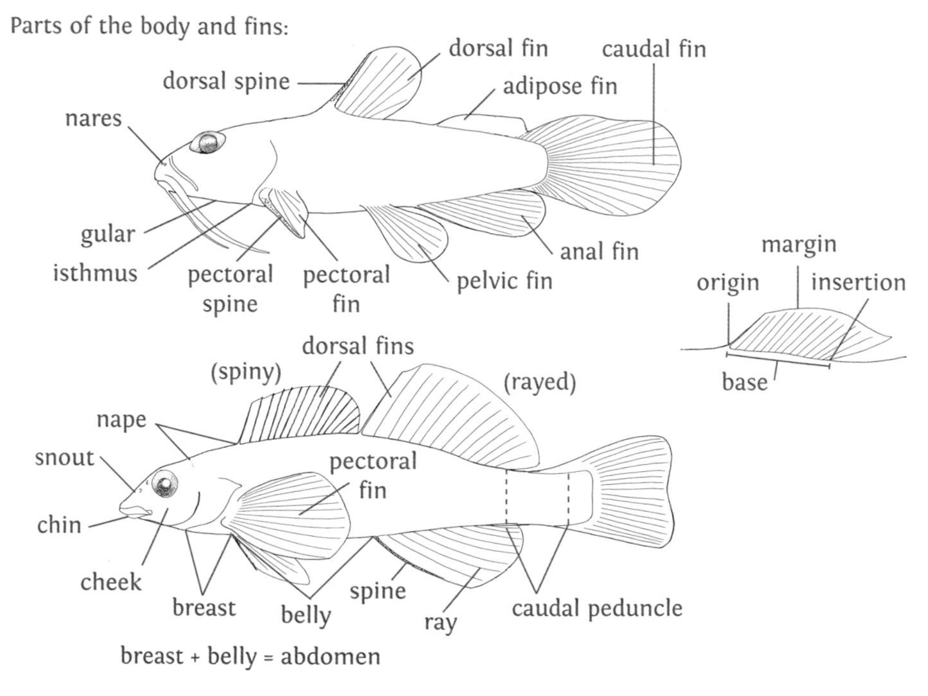

In most fishes, the body is covered with scales. Although some fishes, such as catfishes, may have no scales, some have scales only on a portion of the body. Sturgeons have bony plates, called scutes, arranged in rows along the back and sides. Paddlefish have five types of small scales on different parts of the body but appear scaleless. Gars have thick interlocking rhomboid scales made of a hard mineralized tissue called ganoin—hence the name ganoid scales. Other bony fishes have scales that are thin, overlapping, and flexible. These are smooth cycloid scales in fishes such as minnows, trout, and suckers. There are prickly ctenoid scales in sunfishes, perches, and darters.

General types of fish scales:

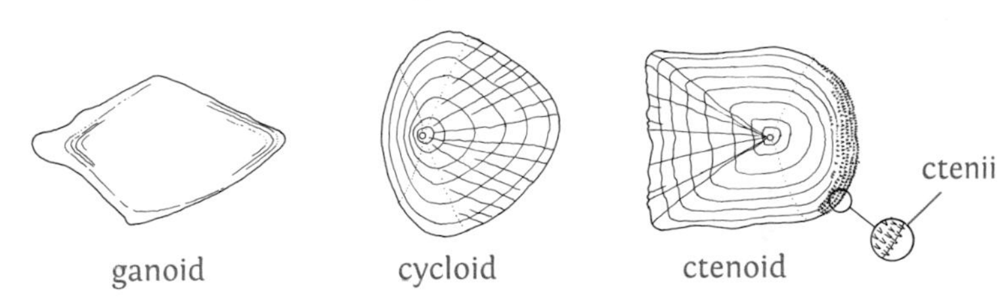

The lateral line is a conspicuous landmark on the body of most fishes. This tubular canal runs along the length of the body at mid-side just underneath the scales. These specialized pores contain sensory nerve endings that allow fishes to detect water vibrations. The visible lateral line and its position may be a diagnostic characteristic for many fishes.

Variations in body form also reveal adaptations or habits for living, feeding, and breeding in special habitats. A fish's form is related to its mode of swimming, which may be specialized for cruising, accelerating, or maneuvering. Even before you identify a specimen, you can presume its likely habits by its body form, position and size of fins, and mouth size and position. Most fishes are propelled by the undulatory motion of the body and the caudal fin. However, fishes that are specialized for cruising have greater muscle mass in the body and a forked caudal fin. These are called rover predators. American Shad and Rosyface Shiner are two examples of fishes specialized for fast swimming. Deep bodied, compressed fishes, such as sunfish, are adapted for maneuvering in complex, vegetated habitats. Pickerels are lie-in-wait predators, specialized for fast acceleration. They have a torpedo-shaped body, forked caudal fin, and dorsal fin far back on the body. Many darters and sculpins are bottom dwellers in fast-flowing habitats. These fishes have large pectoral fins suited for bottom clinging or hiding habits. The eel has another specialized body type, its long snake-like body suited for crawling through crevices along the river bottom. Mouth position provides a hint to the habits of a fish. Fishes with supraterminal or superior mouth position, such as topminnows and mosquitofishes, are adapted to surface feeding.

Most freshwater fishes of Virginia are bony fishes. The lampreys lack true bones. Sturgeons and Paddlefish have cartilaginous skeletons and asymmetric heterocercal caudal fins, where the upper lobe is much larger than the lower lobe. Gar and Bowfin are primitive bony fishes that have caudal fins that are also asymmetric and called abbreviate heterocercal, because the asymmetry is less obvious than in than sturgeons and paddlefish. Bony fishes usually have homocercal, symmetrical caudal fins. The outline of these caudal fins reflects the habits of the species and is defined as rounded, truncate, emarginate, or forked.

Caudal-fin shapes:

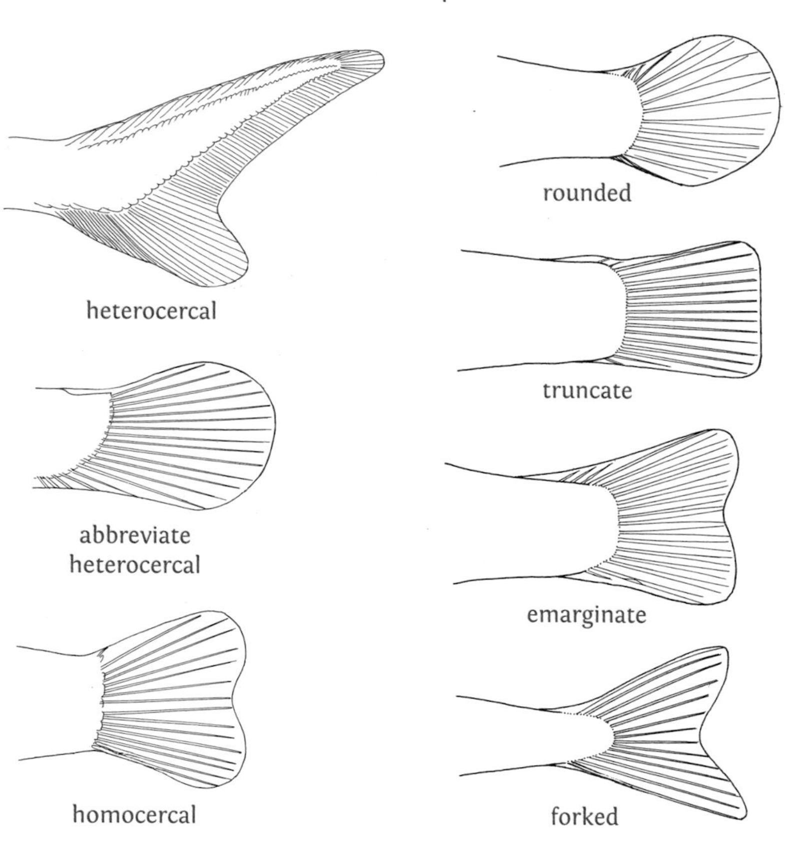

KEY to FRESHWATER FISH FAMILIES of VIRGINIA

1a Paired fins absent; jaws absent; 7 external gill openings, single nostril.................... LAMPREYS—p. 44

1b Paired fins present; jaws present; single external gill opening; paired nostrils.................... 2

2a Tail is asymmetrical, i.e., heterocercal or abbreviated heterocercal shape....... 3

2b Tail is symmetrical, homocercal shape.................... 6

3a Snout is long and paddle shaped; operculum long and flexible; tail deeply forked.................... PADDLEFISH—p. 50

3b Snout lacking paddle-like structure, opercular flap short.................... 4

4a Snout is conical or shovel shaped; 4 long barbels on ventral surface of snout; body with 5 rows of large bony scutes, mouth ventral.................... STURGEONS—p. 47

4b Snout not conical or shovel shaped; without barbels; lacking bony scutes; mouth terminal.................... 5

5a Snout long and beak-like; cylindrical body covered with hard, diamond-shaped scales; dorsal fin far back on body.................... GARS—p. 51

5b Snout rounded and short; body with overlapping cycloid scales; dorsal fin with long base (more than half of body length).................... BOWFINS—p. 52

6a Pelvic fins absent.................... 7

6b Pelvic fins present.................... 8

7a Body very elongated; dorsal, anal, and caudal fins continuous; very small scales (appear scaleless).................... EELS—p. 53

7b Body somewhat elongated; small (< 3 inches); dorsal, anal, and caudal fins separate; caudal fin is rounded; tiny eyes.................... CAVEFISHES—p. 117

8a Adipose fin present.................... 9

8b Adipose fin absent.................... 11

9a Barbels present around mouth; strong sharp spines at front of dorsal and pectoral fins....................................NORTH AMERICAN CATFISHES—p. 100

9b No barbels present around the mouth.. 10

10a Dorsal, pelvic, and anal fins with weak spines; ctenoid scales.............................. .... TROUT-PERCHES—not shown

10b Fins without spines; cycloid scales; axillary process just above pelvic fin........... ....TROUTS—p. 109

11a Belly compressed, keel-like in cross section......................... HERRINGS—p. 54

11b Belly rounded, not keel-like 12

12a Snout duck-bill shaped .... PIKES—p. 112

12b Snout not duck-bill shaped 13

13a Anus located between gill covers anterior to pelvic fins ....PIRATE PERCH—p. 116

13b Anus located posterior to pelvic fins.. 14

14a One dorsal fin present.. 15

14a Two dorsal fins, one spiny and one soft rayed, present 20

15a Long dorsal and anal fins without spines; strong jaw with sharp canine teeth... ....SNAKEHEADS—p. 175

15b Jaw small and without sharp canine teeth... 16

16a Caudal fin forked or emarginated; head without scales 17

16b Caudal fin rounded or truncate 18

17a Mouth supraterminal to inferior without fleshy lips ....MINNOWS and CARPS—p. 60

17b Mouth inferior, with fleshy lips... SUCKERS—p. 90

18a Frenum, bridge of tissue connecting upper jaw to snout, present, which allows the mouth to protrude slightly................................. MUDMINNOWS—p. 115

18b Frenum is absent... 19

19a Dorsal-fin base entirely behind the anal-fin base; third anal-fin ray not branched .... LIVEBEARERS—p. 123

19b Dorsal-fin base almost entirely over anal-fin base; third anal ray branched .... TOPMINNOWS—p. 119

20a Pelvic fin a single prominent spine; first dorsal fin with widely separated spines .... STICKLEBACKS—p. 124

20b Pelvic fin lacking a strong spine 21

21a Dorsal fins well separated; pectoral fin high on body, with base mostly above the lateral midline; pelvic fins in abdominal position; silvery sides and belly .... NEW WORLD SILVERSIDES—p. 118

21b Pelvic fins in thoracic position; dorsal fins joined or only narrowly separated 22

22a Body not scaled; pectoral fins greatly enlarged .... SCULPINS—p. 125

22b Body scaled 23

23a Anal fins with 1–2 spines 24

23b Anal fins with 3 or more spines 25

24a Lateral line extends onto caudal fin; second anal spine much longer than first anal spine .... DRUMS—p. 174

24b Lateral line does not extend onto caudal fin; second anal spine, if present, as long as first anal spine .... PERCHES—p. 151

25a Dorsal fins well connected, with at most a deep notch between; no sharp spine at the edge of the opercle .... SUNFISHES—p. 132

25b Dorsal fins separated by a deep notch; sharp spine near edge of the gill cover .... TEMPERATE BASSES—p. 129

SPECIES ACCOUNTS

Riverweed Darter
Etheostoma podostemone

Candy Darter
Etheostoma osburni

LAMPREYS

Family Petromyzontidae

The eel-shaped lampreys belong to the most primitive vertebrate group in Virginia. Unlike more recognizable fish species, lampreys lack jaws, paired fins, scales, a bony skeleton, and external gill (opercular) covers. Instead, lampreys have an oral disc mouth with horny teeth, a scaleless body, a cartilaginous skeleton, and seven pairs of external gill openings. The tooth pattern of adults is a main method for species identification (see below). All lamprey species spend several years as a non-parasitic larva known as an ammocoete. The blind and toothless larvae live buried within soft sediments of sand and silt feeding on algae, bacteria, and detritus. There are five species found in Virginia and only two of them (Ohio Lamprey and Sea Lamprey) are parasitic as adults. Parasitic species feed on fishes for a year or more before spawning and subsequently dying. The "brook" lampreys are small as adults and do not feed. These non-parasitic species spend several months as non-feeding adults before spawning and subsequently dying.

Lamprey oral discs:

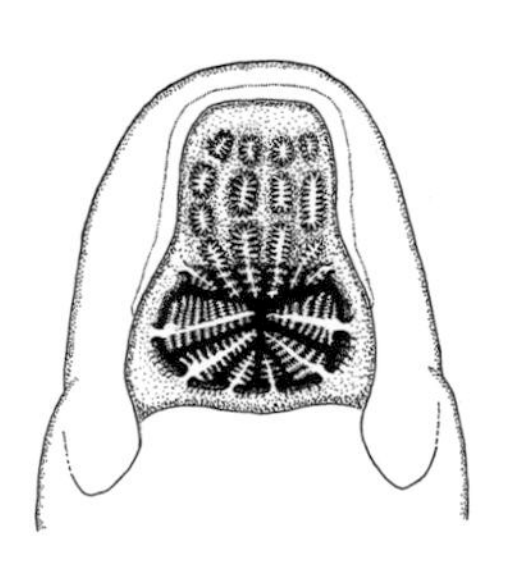

Ammocoete larva

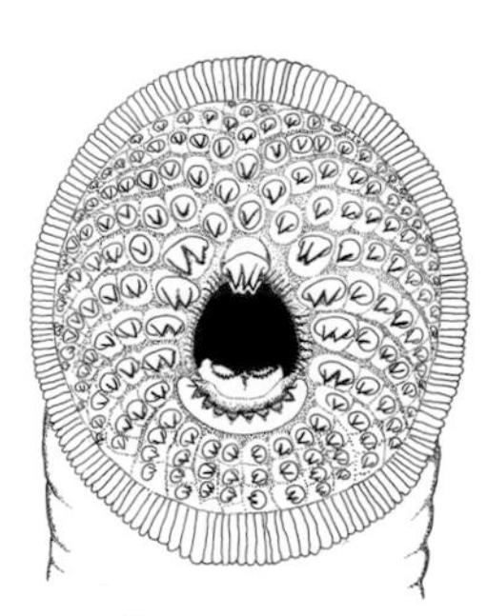

Ohio Lamprey

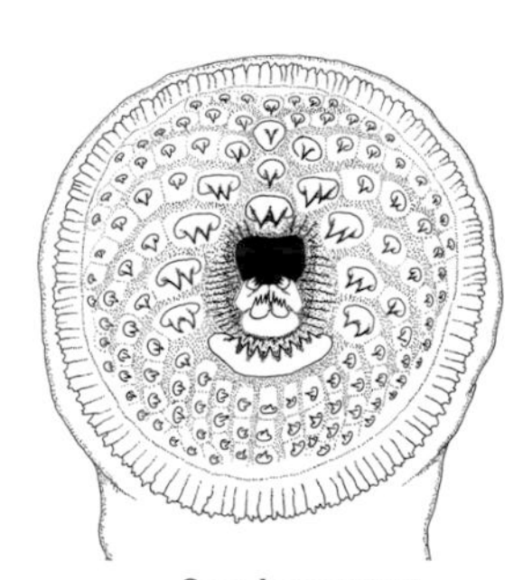

Sea Lamprey

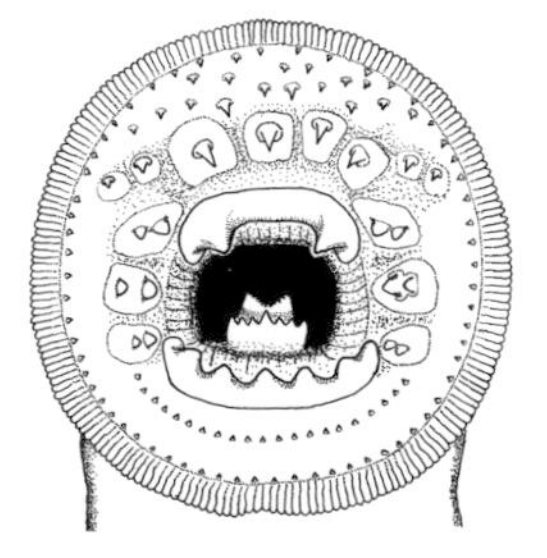

American Brook Lamprey

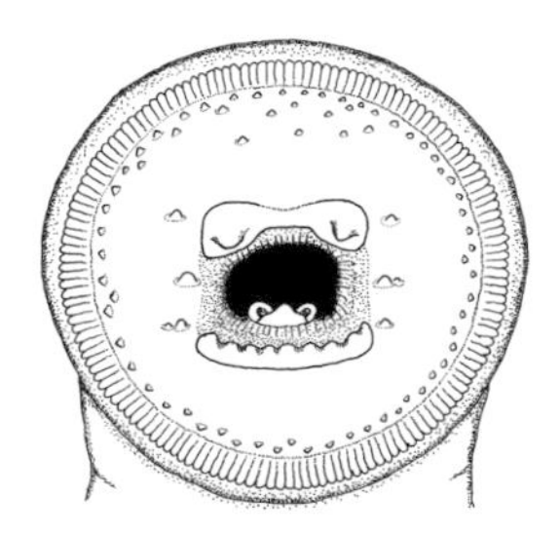

Least Brook Lamprey

Ohio Lamprey

Ichthyomyzon bdellium

Size: 10–14 in.
Habitat: Small and large montane rivers.
Abundance: Uncommon.
Status: Native.

Description: Rounded frontal view; very elongate profile; long curved sharp teeth, oral disc wider than head; single dorsal fin; yellow gold to olive gray above to pale below, dark spots along lateral line. **Reproduction:** Spring spawner; nest builder over sand and pebbles. **Food:** Filter feeds on bacteria, algae, and detritus when larva; parasitic on fishes as adult. **Notes:** Hosts include carp, suckers, black bass, and catfish.

Mountain Brook Lamprey

Ichthyomyzon greeleyi (not shown)

Size: 5–7 in.
Habitat: Small montane creeks and rivers with clean substrate.
Abundance: Uncommon.
Status: Native.

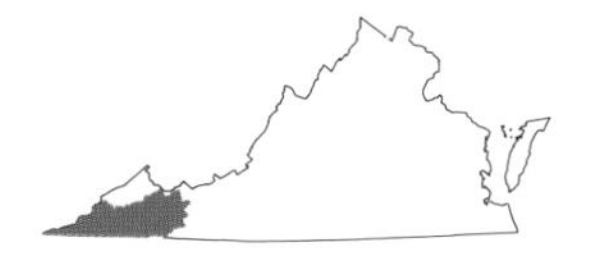

Description: Round frontal view; very elongate profile; short to long straight teeth, oral disc narrower than head; single dorsal fin; black to dark olive with some mottling on sides, dark spots along lateral line, white belly. **Reproduction:** Spawns in spring; creates nest depressions over sand and gravel. **Food:** Non-parasitic larva filter feeds on bacteria, protozoans, detritus, and decayed phytoplankton; adult does not feed. **Notes:** Short lived, five to six years; some populations have possibly been extirpated due to pollution and habitat loss.

Least Brook Lamprey

Lampetra aepyptera

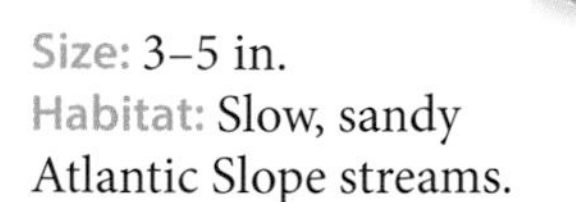

Size: 3–5 in.
Habitat: Slow, sandy Atlantic Slope streams.
Abundance: Uncommon.
Status: Native.

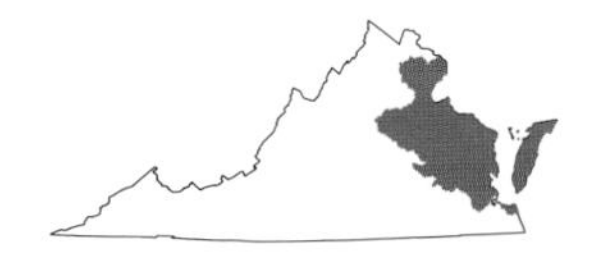

Description: Round frontal view; elongate profile; oral disc with short blunt teeth, oral disc narrower than head; two dorsal fins; tan to gray brown above and on sides, white gray belly. **Reproduction:** Spawns in spring over sand and small gravel. **Food:** Non-parasitic larva filter feeds on bacteria, protozoans, detritus, and algae; adult does not feed. **Notes:** Least Brook Lamprey is the smallest of the lampreys found in Virginia.

American Brook Lamprey

Lethenteron appendix

Size: 5–7 in.
Habitat: Small sandy rivers to slow creeks.
Abundance: Uncommon.
Status: Native.

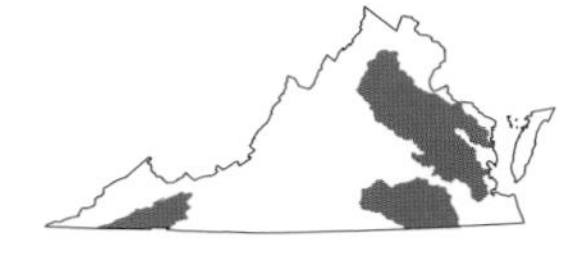

Description: Round frontal view; very elongate profile; oral disc with long rounded teeth, oral disc narrower than head; two dorsal fins; yellow brown above and on sides, white gray belly, dark blotch on caudal fin. Spawning adult is blue black. Reproduction: Spawns in the spring over gravel-sand bottoms and forms local aggregations. Food: Non-parasitic larvae filter feed on bacteria, protozoans, detritus, and algae; adults do not feed. Notes: The species name *appendix* refers to the long genital papilla found on nuptial males.

Sea Lamprey

Petromyzon marinus

Size: 15–25 in.
Habitat: Coastal or Piedmont rivers and estuaries.
Abundance: Uncommon.
Status: Native.

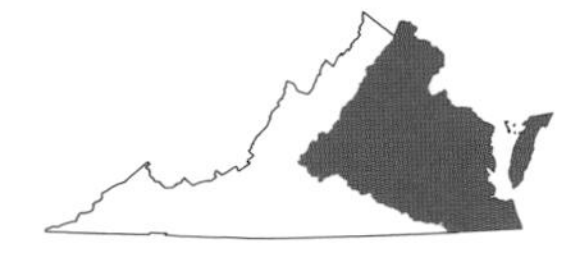

Description: Round frontal view; elongate profile; oral disc with sharp, curved well-developed teeth, oral disc wider than head; two dorsal fins; dark tan to brown with some mottling and white belly. Reproduction: Springtime migration into rivers from the sea or lakes; builds nest in gravel-sand riffles. Food: Filter feeds on bacteria, algae, and detritus when larva; parasitic on fishes as adult. Notes: It feeds on adult fishes by attaching with its teeth, rasping with its tongue, and secreting an anticoagulant called lamphredin.

STURGEONS
Family Acipenseridae

Sturgeons are easily identified by the elongate body with five rows of large bony scutes, four conspicuous barbels in front of a highly extendable, ventral mouth (see below), and asymmetric tail. Worldwide, there are about 27 species in 4 genera, although the evolutionary relationships among species are unclear. However, they are actually ancient bony fishes that lived during the Cretaceous period during the age of dinosaurs (145.5 and 65.5 million years ago). The body form of sturgeons is adapted for cruising along the bottom of rivers, lakes, and oceans, where they locate prey with taste buds on barbels and electroreceptors on the head and snout. Some sturgeon species can live as long as humans do. The largest female sturgeons can produce millions of eggs and do not breed every year. Because of their life history, sturgeons are highly vulnerable to overfishing and dams that block their spawning migrations; therefore, all species worldwide are imperiled in at least portions of their ranges. Two species occur in Virginia waters.

Sturgeon snout, barbels, and mouth:

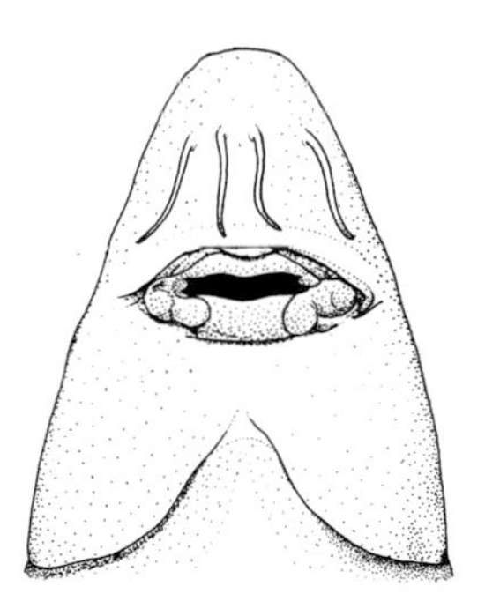

Shortnose Sturgeon

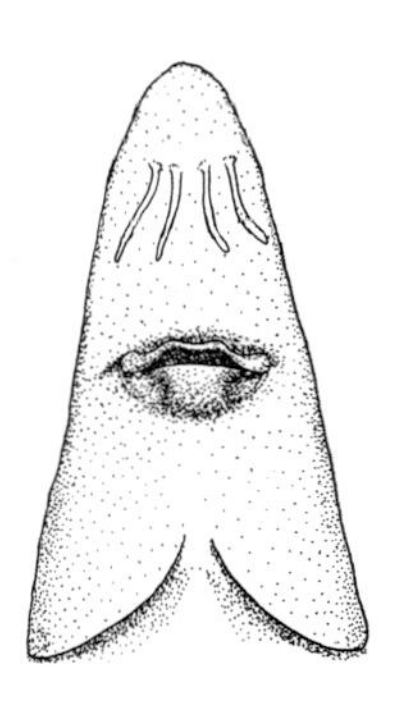

Atlantic Sturgeon

Shortnose Sturgeon

Acipenser brevirostrum

Size: 1.5–4.5 ft.
Habitat: Large rivers and bay estuaries.
Abundance: Rare.
Status: Native.

Description: Rounded frontal view with flat belly; elongate profile; arched back; four barbels on underside of snout; snout short to blunt; inferior mouth; measurement of mouth width inside lips is usually less than 55% of width between eyes; five rows of bony plates on back, side, and abdomen; bluish black above and on upper sides, brown to tan on lower sides, white below; usually no bony plates between anal base and lateral row of scutes. **Reproduction:** Spawns in spring; eggs are laid in fast current over gravel and cobble. **Food:** Benthic invertivore. **Notes:** There are very few documented records of Shortnose Sturgeon in Virginia, with the most recent collected in the James River in 2016. It has been listed as federally endangered since 1967.

Atlantic Sturgeon

Acipenser oxyrinchus

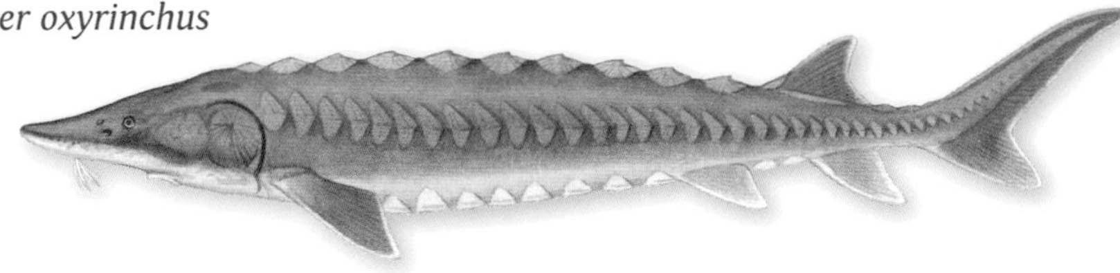

Size: 4–14 ft.
Habitat: Large rivers and bay estuaries.
Abundance: Rare.
Status: Native.

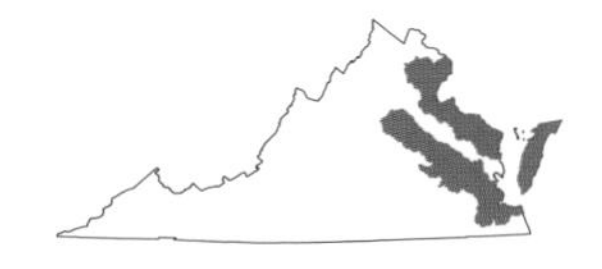

Description: Rounded frontal view with flat belly; elongate profile; arched back; four long barbels; inferior mouth; measurement of mouth width inside lips usually exceeds 62% of width between eyes; snout long to slightly pointed in adult, with tip turned upward in young; fins have pale margins; five rows of bony plates on back, side, and belly; two to six bony plates between anal base and lateral row of scutes; olive green, bluish gray, or brown above and on upper sides, pinkish tan on lower sides, white below.

Atlantic Sturgeon
continued:

Reproduction: In Virginia, spawns in spring and early fall; adhesive eggs are laid over gravel, cobbles, and boulders in tidal freshwaters. Food: Benthic aquatic invertivore. Notes: Despite a coastwide fishing moratorium since 1998, the Atlantic Sturgeon is severely depleted throughout its range. It was listed as federally endangered in the Chesapeake Bay in 2012. A subspecies, the Gulf Sturgeon *Acipenser oxyrinchus desotoi*, occurs in some rivers that drain to the Gulf of Mexico.

A fisheries biologist cradles a juvenile Atlantic Sturgeon.

PADDLEFISH
Family Polyodontidae

The family Polyodontidae is an ancient group that lived over 100 million years ago. Only two species survived to modern times, one in North America and another in China. The Chinese Paddlefish *Psephurus gladius* is likely extinct, a victim of overharvesting, damming, and water pollution. The most notable Paddlefish characteristic is its large oar-like snout. The paddle contains tens of thousands of electrosensory organs that detect weak electric fields created by zooplankton. Instead of a bony, calcified skeleton, the skeleton of a Paddlefish is composed mostly of cartilage. Paddlefish feed by opening their large mouth and filtering food from the water column. The family and genus names are derived from the Greek *poly*, meaning many, and *odon*, meaning tooth, referring to the many gill rakers essential for filter feeding. Paddlefish can be found in large rivers and reservoirs in the Mississippi River drainage, although the historic range has shrunk.

Paddlefish
Polyodon spathula

Size: 3.5–4.5 ft.
Habitat: Warm medium to large rivers.
Abundance: Rare.
Status: Native.
Other names: Spoonbill, Spoonbill Cat.

Description: Rounded frontal view; elongate profile; large, flat paddle-shaped snout protruding from huge mouth; scaleless body; heterocercal caudal fin; very small eye; black to gray above and on sides, white below. Reproduction: Spawns early spring to early summer; a female may lay over 1 million eggs over flooded gravel bars. Food: Plankton and small aquatic invertebrates. Notes: The Paddlefish is protected in Virginia as a threatened species. In other areas of the United States, it is harvested by commercial fishing and recreational snagging. Paddlefish is prized for its meat and roe, the latter marketed as caviar.

GARS
Family Lepisosteidae

Gars are easily identified by their elongate, torpedo-shaped bodies, hard rhomboid scales, and long snouts lined with sharply pointed teeth. There are seven species in this ancient group of bony fishes, five in eastern North America and one each in Cuba and Central America. Living gars are in two genera, *Lepisosteus* and *Atractosteus*. Gars are ancient fishes, with relatives dating back 100 million years or more. Their closest relatives are extinct gars; some had sharp pointed teeth on long snouts and some had flattened teeth on short snouts for crushing mollusks. Gars can breathe via gills, although in low-oxygen conditions they can breathe air with a lung-like swim bladder. They are mistakenly thought to decimate game fish populations but instead feed on small fishes, making them an important component of a healthy aquatic environment. Gars are long lived with high fecundity, and their yolk sac larvae attach to vegetation with an adhesive organ. Their eggs are poisonous if eaten.

Longnose Gar
Lepisosteus osseus

Size: 20 in. to 3.75 ft.
Habitat: Warm, medium to large rivers and reservoirs.
Abundance: Common.
Status: Native.
Other name: Garpike, Needlenose Gar.

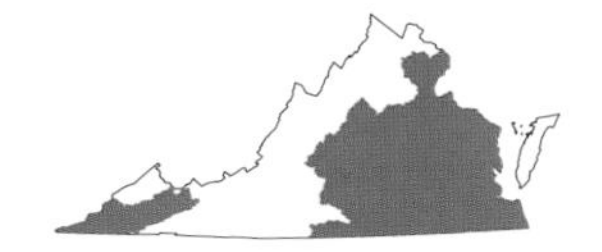

Description: Round frontal view; elongate profile; spindle-shaped body; long snout with many sharp, needle-like teeth; plate-like, rhomboid scales; color overall olive or silvery, often with dark spots, especially toward tail; fins spotted. Reproduction: Spawns in spring, broadcasting adhesive eggs over various substrates, including dense vegetation. Food: Small fishes, amphibians, and aquatic invertebrates. Notes: Longnose Gar may live to 32 years. It may also be encountered throughout estuaries of Virginia. Its large size and near-surface habits make Longnose Gar vulnerable to bow-fishers. Meat is well flavored.

BOWFINS
Family Amiidae

Bowfins are primitive bony fishes restricted to lowland habitats of eastern North America. A single species, *Amia calva*, is present today, although extinct members of this family were described from fossil specimens present in sediments dated from 40 to 150 million years before the present. In prehistoric times, bowfins were distributed on all continents except Australia and Antarctica. Extinct species were similar to present day Bowfin, with a long, nearly cylindrical body, abbreviate heterocercal tail, long dorsal fin, tubular nostrils, functional lung, and large bony plate on the throat. Bowfin can gulp air when water temperature is warm (86°F) and oxygen levels are low. Like gar, Bowfin are an important component of a healthy aquatic environment.

Bowfin
Amia calva

Size: 2–3 ft.
Habitat: Slow, stagnant rivers, lakes, and estuaries.
Abundance: Common.
Status: Native.
Other names: Grindle, Blackfish, Dogfish, Choupique (pronounced "shoe pick"), Mudfish, Cypress Trout

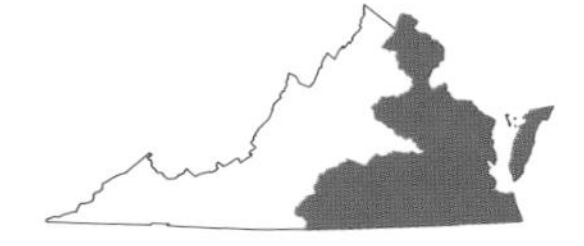

Description: Rounded frontal view; elongate profile; large bony head with tubular nostrils; large mouth with row of short, sharp teeth; long uniform dorsal fin; caudal fin is rounded, with a dark spot; base color olive to tan above and sides mottled in dark olive, white to pale below; head has three dark bars; breeding male has turquoise blue green lips, mouth, and fins and an orange halo around the caudal-fin spot. Reproduction: Spawns in spring; male constructs a bowl-shaped nest among vegetation and other cover and aggressively defends eggs and young. Food: Ambush predator on fishes, amphibians, and aquatic invertebrates. Notes: Although not a food fish, Bowfin is noted as a good fighter and was once stocked outside its native range for angling opportunities. Bowfin is sometimes confused with Northern Snakehead (see p. 175), which has a much longer anal fin and a lower jaw that protrudes past the upper jaw.

FRESHWATER EELS
Family Anguillidae

The family Anguillidae consists of 17 species worldwide that share a unique life cycle. They begin life in the ocean, then migrate to spend most of their life in fresh or brackish water before returning as mature adults to spawn at sea. This pattern of migration is known as catadromy, meaning "down running." American and European freshwater eels spawn and die soon after at great depths in distinct portions of the Sargasso Sea. After hatching from buoyant eggs, the ribbon-shaped, transparent American Eel larvae known as "leptocephali" drift with ocean currents to coastal habitats from Suriname to Iceland. As they approach the coast, they metamorphose to shorter, semi-pigmented glass eels, which continue migrating to nearshore habitats and fresh waters. In estuaries and river mouths, morphology and coloration change again as the glass eels become pigmented elvers and ultimately yellow eels. Freshwater eels are long lived (up to 20+ years in fresh water). Fecund fish and females grow larger than males. Their bodies swim by undulating and are specialized for entering crevices.

American Eel
Anguilla rostrata

Size: 9 in. to 2.6 ft.
Habitat: Warm, medium to large rivers and reservoirs.
Abundance: Common.
Status: Native.
Other names: Black Eel, Bronze Eel, Glass Eel, Elver, Silver Eel, Yellow Eel

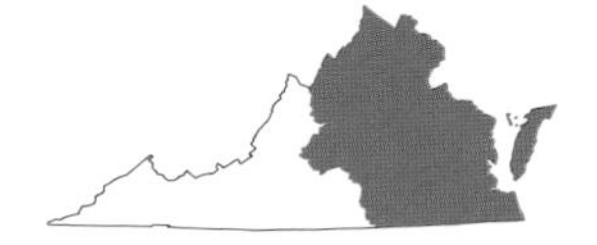

Description: Round frontal view, compressed in posterior half; very elongated profile; dorsal and anal fins are long and continuous with the caudal fin; large mouth, with end of jaw extending past eye; juvenile olive to brown above and on side, with yellow or white belly; breeding adult black above, dark bronze upper, and silver sides and below; black pectoral and caudal fin in mature adult. **Reproduction:** Spawns in fall to early winter; female produces 0.5 to 2.5 million eggs. **Food:** Highly dependent on availability but includes a wide variety of aquatic invertebrates, fishes, and carrion. Eels in turn are important food for many species. **Notes:** American Eels are harvested as a food fish at larger sizes, particularly for sushi (but not raw) or hot smoking. Elvers are harvested for bait, and glass eels are highly valued by the aquaculture industry for grow-out. It is illegal to harvest glass eels from Virginia waters.

HERRINGS
Family Clupeidae

Clupeid fishes are adapted for life in well-lit pelagic zones of lakes, rivers, and the ocean. All 195 species of the Clupeidae are silvery with a green blue dorsum, laterally compressed bodies, and sawblade-like scutes on their bellies. Bait dealers often sell them as "sawbellies." The combination of counter-shaded coloration, silvery mirror scales, and schooling behavior are effective anti-predator strategies. Herrings have sensitive hearing, and some species of the genus *Alosa* can detect ultrasound. In the fresh waters of Virginia, you may encounter six or seven clupeid fishes. Four are anadromous species that spend most of their life cycle in salt water, two are freshwater, and the other is euryhaline. All herrings are highly prolific yet fragile fishes, and their abundance may change quickly due to die-offs or recruitment booms. The anadromous species are particularly important because they transfer marine-derived nutrients to fresh water in addition to serving as prey for numerous coastal birds, mammals, and fishes. Atlantic Menhaden, *Brevoortia tyrannus*, is an abundant clupeid in Chesapeake Bay and coastal waters and sometimes enters tidal freshwaters. Blueback Herring, *Alosa aestivalis*, and Alewife, *Alosa pseudoharengus*, look very similar and are often referred to collectively as river herrings. American Shad, *Alosa sapidissima*, is the largest and best-known anadromous herring. Gizzard shads (genus *Dorosoma*) are named for a distinctive gizzard-like stomach, and the last dorsal-fin ray is an elongated filament. There are two species of *Dorosoma* in Virginia, the Gizzard Shad, *Dorosoma cepedianum*, and the Threadfin Shad, *Dorosoma petenense*.

Snout and jaw shapes of herrings:

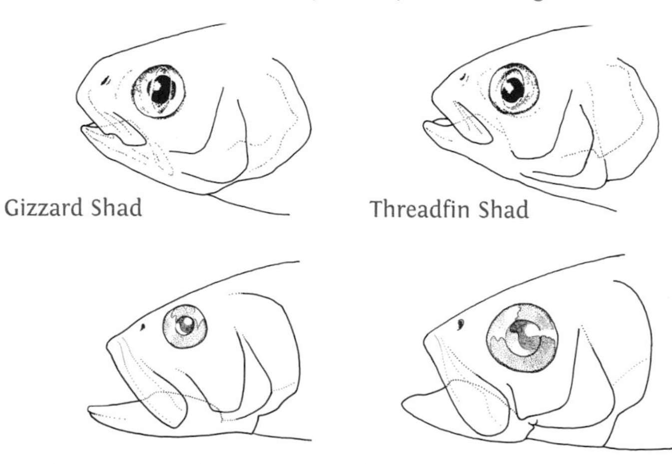

Gizzard Shad

Dorosoma cepedianum

Size: 7–13 in.
Habitat: Low-gradient rivers, reservoirs, lakes, swamps, and estuaries.
Abundance: Common.
Status: Native.

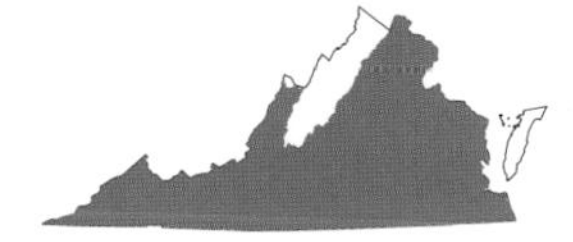

Description: Strongly compressed frontal view; oval profile; blunt snout; tip of snout extends beyond the tip of lower jaw; deep notch on the center of upper jaw; small subterminal mouth; long, whip-like last dorsal ray; dark shoulder spot is most prominent in juveniles and fades with age; silvery green coloration above, fading to plain silver below; often with a series of darkened scales forming stripes on the dorsum. Reproduction: Large spawning aggregations occur in spring; breeding individuals swim at the surface, and eggs and sperm are scattered. Eggs adhere to the bottom. Food: Filter feeds on phytoplankton when young and switches to zooplankton as it grows larger. Notes: Similar to Threadfin Shad, which has more pointed snout that does not extend beyond the tip of lower jaw and lacks a notch on lower jaw (see p. 54). Threadfin Shad has fewer anal-fin rays (20–25) than Gizzard Shad (29–35). Introduced widely in reservoirs for forage base for game fishes. Caught with cast nets or dip nets and used for bait. Bald Eagles, Ospreys, and Great Blue Herons can capture a large Gizzard Shad.

Threadfin Shad

Dorosoma petenense

Size: 5–8 in.
Habitat: Low-gradient rivers, reservoirs, lakes, swamps, and estuaries.
Abundance: Common.
Status: Introduced.
Other name: Hickory Shad.

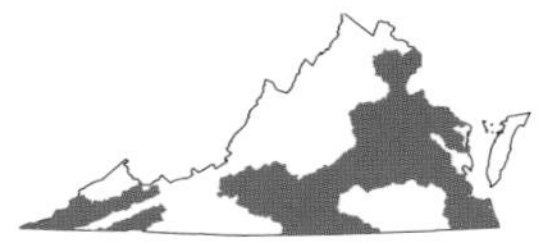

Description: Strongly compressed frontal view, oval profile; pointed snout and small terminal mouth; snout tip does not extend beyond tip of lower jaw; lacks a notch on lower jaw; long, whip-like last dorsal ray; dark shoulder spot is most prominent in juveniles and fades with age; silvery green coloration above, fading to plain silver below with some yellow on fins, particularly the tail. Reproduction: Large spawning aggregations occur in spring; breeding individuals swim at the surface, and eggs and sperm are scattered. Eggs adhere to the bottom. Food: Filter feeds on phytoplankton when young and switches to zooplankton as it grows larger. Notes: Similar to Gizzard Shad, which has a blunter snout, its tip extends beyond the tip of lower jaw and has a notch on the ventral margin of upper jaw (see p. 54). Caught with cast nets or dip nets and used for bait. Prone to winterkill in Virginia.

Blueback Herring

Alosa aestivalis

Size: 7–15 in.
Habitat: Tidal and nontidal rivers and creeks.
Abundance: Locally abundant.
Status: Native.

Description: Strongly compressed frontal view, deep profile with elliptical shape; oblique mouth with small teeth on jaws; eye diameter about equal to snout length; cheek length wider than its depth (see p. 54); typically, one blue-black shoulder spot near upper gill-cover edge; back and upper back are blue with thin dusky-black stripes; lower sides silver.

Blueback Herring
continued:

Other names: Blueback Shad, River Herring.

Reproduction: Anadromous. Scatters eggs on riverbed in March and April in tidal and nontidal rivers and creeks. Food: Mostly planktivorous. Notes: A landlocked population exists in Kerr Reservoir. Very similar in appearance to the Alewife, *Alosa pseudoharengus*, which is known to hybridize with Blueback Herring. Back of Alewife is typically gray green, whereas the Blueback Herring is blue green. Alewife has larger eye and pale gray or pinkish white lining inside the belly, whereas the belly lining of the Blueback Herring is sooty or black. Harvest moratorium on river herrings has been in effect since 2012.

Alewife

Alosa pseudoharengus

Size: 7–14 in.
Habitat: Tidal rivers and creeks and reservoirs.
Abundance: Locally abundant.
Status: Native.
Other names: Gaspereau, Branch Herring, River Herring.

Description: Strongly compressed frontal view, deep profile with ellipical shape; oblique mouth with teeth on jaws; eyes large; cheek length about equal to its depth (see p. 54); back and upper back are green gray, with thin dusky-black stripes; typically, one blue-black shoulder spot near upper gill-cover edge; lower sides are silver. Reproduction: Anadromous. Scatters eggs on riverbed in February to April in tidal and nontidal rivers and creeks. Juveniles migrate downstream. Food: Mostly planktivorous. Notes: Landlocked populations exist in Claytor Lake, Lake Moomaw, Flannagan Reservoir, and Smith Mountain Lake. Back of Alewife is typically gray green, whereas the Blueback Herring is blue green. Alewife has larger eye and pale gray or pinkish white belly lining, whereas the belly lining of the Blueback Herring is sooty or black. The harvest moratorium on river herrings has been in effect since 2012.

Hickory Shad

Alosa mediocris

Size: 15–22 in.
Habitat: Tidal rivers.
Abundance: Common.
Status: Native.
Other name: Silver Shad.

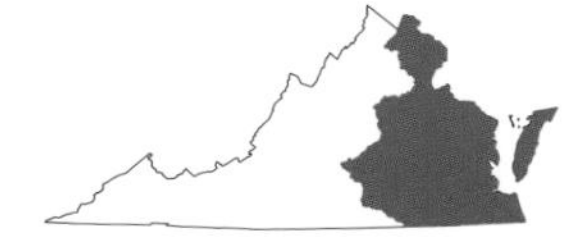

Description: Strongly compressed frontal view, deep profile with elliptical shape; oblique mouth with teeth on jaws; cheek length about equal to its depth (see p. 54); blue green on the back and upper sides shades to silver on the lower sides; a blue-black spot near the upper gill-cover edge is followed by a row of poorly defined dusky spots reaching to below the dorsal fin; small blue spots develop along the rows of scales on adults. Lower jaw projects beyond the snout; fewer (18–23) gill rakers on first arch than river herrings or American Shad. Reproduction: Spawns in tidal freshwater rivers. Scatters eggs. Food: Adult feeds mainly on fishes, small crabs, squids, fish eggs, and pelagic crustaceans. Notes: Similar in appearance to the American Shad and river herrings, except Hickory Shad lower jaw projects beyond the snout (see p. 54). The species name, *mediocris*, translates to "mediocre," referring to its desirability as a food fish.

American Shad

Alosa sapidissima

Size: 20–26 in.
Habitat: Tidal rivers.
Abundance: Uncommon.
Status: Native.
Other name: White Shad.

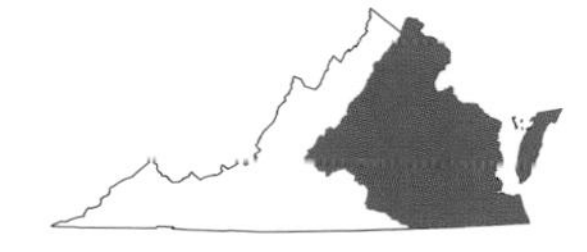

Description: Strongly compressed frontal view, deep profile with elliptical shape; oblique mouth with teeth on jaws; cheek deeper than wide (see p. 54); back and upper sides are green or blue and lower sides are silver; a blue-black spot near the upper gill cover is followed by row of smaller spots; lower jaw is equal to length of or projects only slightly beyond the snout. Reproduction: Spawns in tidal fresh water over shallow flats and riffles, March to April. Fertilized eggs drift in the current until they sink and adhere to sand, gravel, and cobble streambed. In Virginia, spawners return to ocean to migrate and return to spawn again. Food: Planktivorous on a variety of small prey types. Adults do not feed in fresh water. Notes: Species name *sapidissima* means "delicious to eat." Similar in appearance to the Hickory Shad, but Hickory Shad lower jaw projects beyond the snout. Harvest moratorium in all Virginia waters. Once supported a large seasonal commercial fishery. George Washington fished for American Shad in the Potomac River. During the Revolutionary War, Washington and his troops were spending winter near the Schuylkill River at Valley Forge, Pennsylvania, and replenished their food supplies with an early run of American Shad. In Virginia and northern populations, adults return to spawn multiple times; however, in southern populations adults breed once and die.

MINNOWS and CARPS
Family Cyprinidae

Cyprinidae is among the largest families of freshwater fishes worldwide, with at least 1,675 known species. These carps, minnows, dace, shiners, and chubs are often and incorrectly characterized as small and insignificant. Nothing could be further from the truth. Common Carp and Grass Carp are large minnows that can reach over 3 feet in length, Fallfish, or "Shenandoah tarpon," may reach 18 inches, and there are even larger species in Southeast Asia. Cyprinids play an essential role in the ecosystem by moving energy through the food web from smaller to larger organisms. Cyprinids do not have teeth on the jaws like many fishes, but instead they are located in the throat, attached to the interior gill arches. The family is characterized by a single dorsal fin (lacking a spine in native species), smooth rounded scales, a complete lateral line, and pelvic fins on the abdomen. Most cyprinids are found in small streams to large rivers of the montane region and Piedmont, although a few species are adapted to the swampy areas of the Coastal Plain. Across North America, 46% of the species in this family are imperiled. Taxonomists recently elevated the subfamily Leuciscinae to family status, so recent references may label North American cyprinids as Leuciscidae.

Mouth and head shapes of cyprinids:

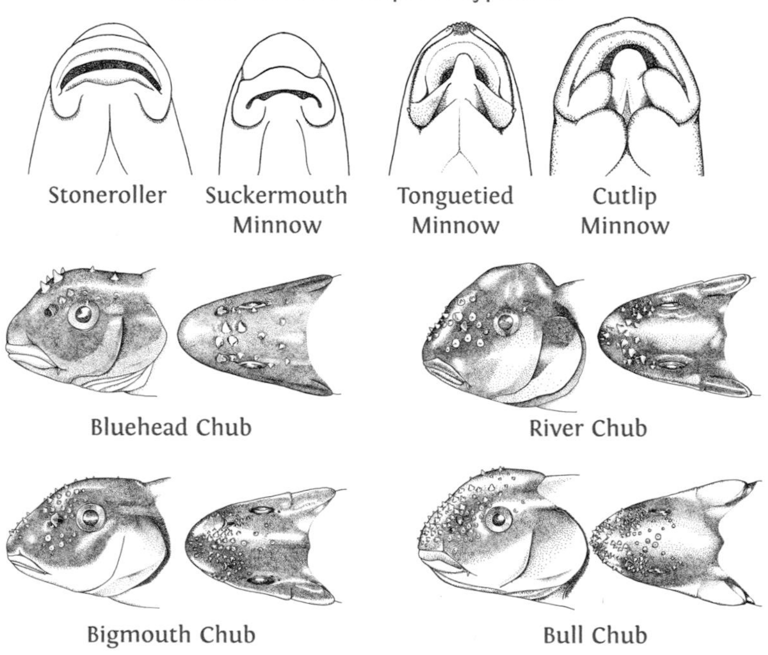

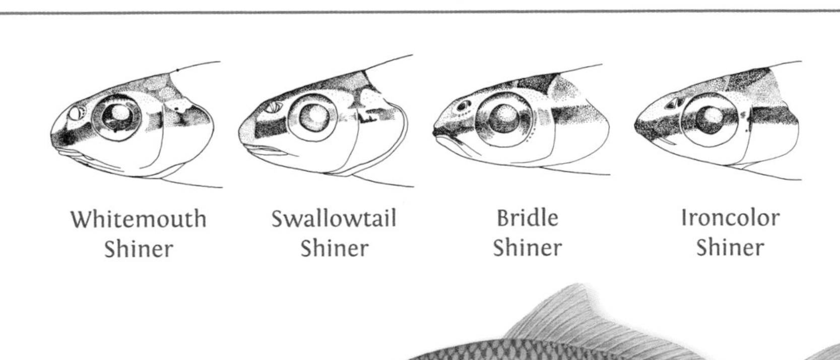

Common Carp

Cyprinus carpio

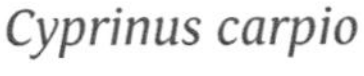

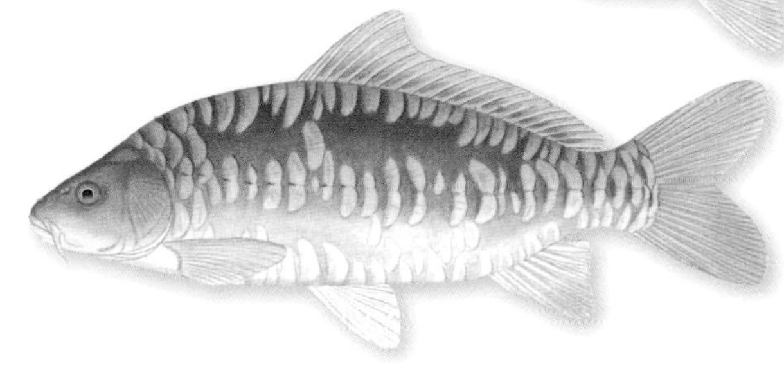

"Mirror" variant

Size: 12 in. to 3 ft.
Habitat: Warm medium streams to large rivers, ponds, and reservoirs.
Abundance: Common.
Status: Introduced.
Other names: Israeli Carp, Koi, Leather Carp, Mirror Carp, Mud Bass.

Description: Rounded to slightly compressed frontal view, deep profile; subterminal mouth; large downsloping head, two long barbels on each corner of jaw; long dorsal fin; a single hard spine on dorsal and anal fins; brassy green above, brassy olive on side, silver to brass below; adult with reddish caudal and anal fins. Cultured ornamental varieties (Koi carp) have a variety of colors, including orange, red, yellow, and black. Some varieties nearly scaleless (leather carp) or having a few enlarged scales (mirror carp). **Reproduction:** Spawns in spring to early summer; mating occurs in backwaters of sluggish rivers and coves of reservoirs. **Food:** Aquatic and terrestrial invertebrates, plants, and detritus. **Notes:** Common Carp was first introduced from Eurasia in the early 1800s. Some anglers actively seek carp for their difficulty to catch and fighting ability. They are a favored food fish in some cultures, especially in Europe.

Goldfish

Carassius auratus

Size: 10–16 in.
Habitat: Cool to warm medium streams to large rivers, ponds, and reservoirs.
Abundance: Uncommon.
Status: Introduced.

Description: Slightly compressed frontal view, deep profile; subterminal mouth; no barbels; large, shimmering scales; long dorsal fin; male has longer pectoral and pelvic fins than female; color varies from brown, olive, silver (wild type coloration) to yellow, white, and orange. Reproduction: Spawns in spring to summer; eggs laid in slow sluggish water on aquatic plants and submerged wood. Food: Algae, aquatic plants, insects, and small fish. Notes: Goldfish was the first introduced exotic fish species in North America. Although a popular ornamental species for backyard ponds and aquaria, Goldfish is scattered throughout Virginia due to illegal releases. It was formerly sold for bait as "Baltimore Minnows" in some areas, and this likely explains the introductions.

Grass Carp

Ctenopharyngodon idella

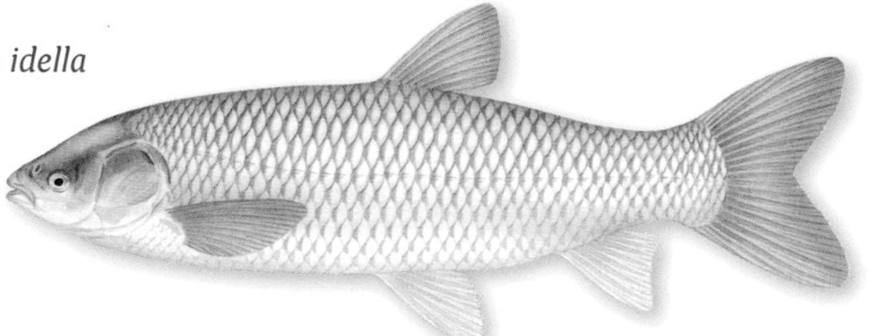

Size: 18 in. to 3.5 ft.
Habitat: Warm ponds, reservoirs, sluggish streams, and rivers.
Abundance: Rare.
Status: Introduced.

Description: Rounded frontal view, elongate to moderate profile; terminal mouth; wide head; deep caudal peduncle; black to olive brown above, side silver, white to yellow below; scales have dark outlines; male and female appear similar.

Grass Carp
continued:

Reproduction: Spawns in spring; several males mate with a single female. Food: Plankton as young; aquatic plants as adults. Notes: Grass Carp was introduced to control excessive vegetation and often escapes ponds and lakes during flooding. All Grass Carp stocked in Virginia waters must be sterile triploids obtained from approved vendors to prevent establishment of reproducing populations.

Golden Shiner

Notemigonus crysoleucas

Size: 3–8 in.
Habitat: Warm, slow-moving streams, rivers, swamps, ponds, and lakes.
Abundance: Common.
Status: Native.

Description: Strongly compressed frontal view, deep profile; small, upturned mouth; highly down-curved lateral line; falcate anal fin; dusky silver to gold above, silver to gold on side, silver to white below; young with dark stripe on side; breeding male with light to bright yellow (sometimes red) fins. Reproduction: Spawns in spring to fall; several males spawn with a female over open water. Food: Aquatic and terrestrial insects, mollusks, and algae. Notes: Golden Shiner has been introduced throughout Virginia, as it is popular for commercial bait production. It is one of the few minnow species that can tolerate low-pH environments.

Mountain Redbelly Dace

Chrosomus oreas (right)

Tennessee Dace

Chrosomus tennesseensis (not shown)

Size: 1¾–2¾ in.
Habitat: Cool to warm small to medium streams.
Abundance: Rare to abundant.
Status: Native.

■ Mountain Redbelly Dace
■ Tennessee Dace

Description: Rounded frontal view, moderate profile; subterminal mouth; tiny scales; breeding males vivid silver, gold, and gray above and on side, lower side and below red; broken stripe on side; small specks (Tennessee Dace) to large (Mountain Redbelly Dace) blotches along upper side and back; pectoral and pelvic fins bright yellow; female less colorful than male; juvenile drab. **Reproduction:** Spawn in spring to early summer; use gravel mound nests of chubs and gravel pits of stonerollers. **Food:** Detritus, algae, and small invertebrates. **Notes:** Mountain Redbelly Dace was introduced into the Holston River system, where it has outcompeted Tennessee Dace, a state endangered species. Members of this genus are very colorful and often used as bait, which may explain expanding ranges.

Clinch Dace

Chrosomus sp. cf. *saylori* (right)

Blackside Dace

Chrosomus cumberlandensis (not shown)

Size: 1¾–2¾ in.
Habitat: Cool to warm small streams.
Abundance: Rare.
Status: Native (Clinch) and introduced (Blackside).

Description: Rounded frontal view, moderate profile; subterminal mouth; tiny scales; breeding males with single, wide black lateral stripe (Blackside Dace) or a single bright-yellow lateral stripe between two narrow black stripes (Clinch Dace); light brown to gold above and upper side, red on lower

Clinch Dace
Blackside Dace
continued:

side and below; small specks (Blackside Dace) to small blotches (Clinch Dace) on upper side and above; pectoral and pelvic fins bright yellow; female less colorful than male; juvenile drab; nonbreeding Blackside Dace has two dark stripes that conjoin near the tail. Reproduction: Spawn in spring to early summer; use gravel mound nests of chubs and gravel pits of stonerollers. Food: Detritus, algae, and small invertebrates. Notes: The undescribed Clinch Dace was recently discovered in the upper Clinch River system. Blackside Dace, a federally threatened species, was introduced into the Powell and Clinch River watersheds.

Rosyside Dace

Clinostomus funduloides

Size: 2–4½ in.
Habitat: Cool to warm small to medium streams.
Abundance: Common.
Status: Native.

Description: Compressed frontal view, moderate profile; large, terminal mouth; olive above, bright red to orange on side, silver below; fins olive yellow; nonbreeding individuals with light-red humeral bar; breeding male with small tubercles over most of body and head; female lacks tubercles, and red is less intense. Reproduction: Spawns in spring and early summer; spawns on chub and Fallfish nests. Food: Aquatic and terrestrial insects. Notes: Rosyside Dace is an indicator of excellent water quality. It uses its large mouth to consume insects from the water's surface and is sometimes caught by fly anglers.

Longnose Dace

Rhinichthys cataractae

Size: 2½–5 in.
Habitat: Cool to cold clear medium to large streams.
Abundance: Common.
Status: Native.

Description: Rounded frontal view, elongate profile; long, downturned snout; subterminal mouth; small barbel in corner of mouth; small scales; small eye; dark lateral stripe beginning at snout and ending in triangular spot before caudal fin, dark olive to reddish purple above, mottled black-brown scales on side, and silver yellow below; fins red in breeding males. Reproduction: Spawns in spring and early summer; eggs laid on gravel bottom in swift areas. Food: Aquatic invertebrates, plants, and detritus. Notes: The reduced swim bladder and streamlined body, adaptations usually reserved for darters and sculpins, allow this species to live in fast currents.

Western Blacknose Dace

Rhinichthys obtusus (right)

Eastern Blacknose Dace

Rhinichthys atratulus
(not shown)

Size: 1¾–4 in.
Habitat: Cold to warm small to medium streams.
Abundance: Common.
Status: Native.

■ Western Blacknose Dace
■ Eastern Blacknose Dace

Description: Rounded frontal view, moderate profile; small scales; long pointed snout; subterminal mouth; small barbel in corner of mouth; brown to olive above, stripe on side black to brown to red (narrow and intense in Eastern, broader and pinkish in Western), silver to white below; fins pale; breeding male with few (Eastern Blacknose Dace) to many (Western Blacknose Dace) scattered black scales on back and side.

Western Blacknose Dace
Eastern Blacknose Dace
continued:

Reproduction: Spawn in late spring to summer; eggs laid on gravel and sometimes nests of chubs and suckers. Food: Aquatic and terrestrial invertebrates, algae, and detritus. Notes: Blacknose Dace is often the only fish species in many small headwater streams.

Central Stoneroller

Campostoma anomalum (right)

Largescale Stoneroller

Campostoma oligolepis (not shown)

Size: 4–7 in.
Habitat: Cool to warm small streams to medium rivers.
Abundance: Abundant.
Status: Native.
Other names: Hornyhead, Knottyhead.

■ Central Stoneroller
■ Largescale Stoneroller

Description: Rounded frontal view, moderate profile; inferior mouth; hard horseshoe-shaped ridge on lower jaw (see p. 60); olive to brown with brassy sheen above, brassy to orange on side, silvery white below; fins olive to pale; breeding male with black band on dorsal fin, dusky spot on caudal fin, orange to pale-yellow fin; anal fin of breeding male with bold dark or faint band (Central Stoneroller), band faint or absent (Largescale Stoneroller); tubercles over entire body, with large tubercles present (Central Stoneroller) to absent (Largescale Stoneroller) on inside portion of nostril; female and juvenile lack tubercles and banding. Reproduction: Spawn in spring; males create stone pits in runs and glides. Food: Aquatic invertebrates, detritus, and algae. Notes: The name "stoneroller" derives from the behavior of turning over rocks to graze on algae and build nest pits. Distinguishing the two species is easiest by examining tubercle arrangement and anal-fin coloration in a breeding male.

Allegheny Pearl Dace

Margariscus margarita

Size: 2–3½ in.
Habitat: Cool to cold small to medium spring-fed streams.
Abundance: Uncommon.
Status: Native.

Description: Rounded frontal view, moderate profile; rounded snout; small scales; dark stripe on back, dark olive to gray above, silver with dark stripe and speckling on side, white below; breeding male with rosy to pink on lower side. Reproduction: Spawns in spring; breeding male develops territory over gravel-sand substrate. Food: Aquatic insects, small mollusks, small fishes, and plants. Notes: The Shenandoah River is the southern extent of this northern species' range.

Fallfish

Semotilus corporalis

Size: 8–18 in.
Habitat: Cool to warm small to large rivers.
Abundance: Uncommon.
Status: Native.
Other name: Shenandoah Tarpon.

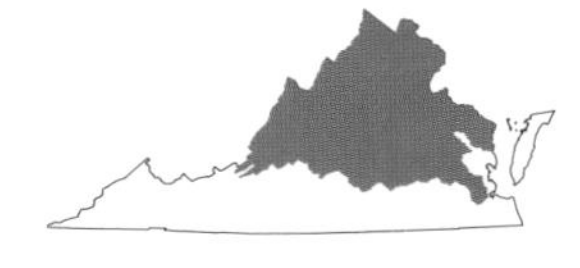

Description: Slightly compressed frontal view, moderate profile; rounded snout; slightly subterminal mouth; front edge of dorsal and pelvic fins nearly aligned vertically; dark lateral stripe with spot before caudal fin in young; adult with darkly outlined scales on upper side and back; olive to brown above; dusky above stripe, silver white below; breeding male has turbercles on head and snout. Reproduction: Spawns in spring; male constructs mound nest of pebbles and small cobbles. Food: Aquatic and terrestrial invertebrates, small fishes, plants, and detritus. Notes: Fallfish is the largest native minnow in the eastern United States. The James River is the southern range limit. The nesting mound of Fallfish can be over 3 ft. in diameter and 1 ft. high.

Creek Chub

Semotilus atromaculatus

Size: 3–7¾ in.
Habitat: Cool to warm small streams to medium rivers.
Abundance: Common.
Status: Native.

Description: Rounded frontal view, moderate profile; rounded snout; front edge of dorsal fin well behind front edge of pelvic fin; dark spot at front base of dorsal fin; gray olive above, dark stripe on sides, silver on side with brassy to violet shimmering scales, white belly; breeding adult colored pink on lower body; blue on side of head; ventral fins orange; three to seven large tubercles on head. Reproduction: Spawns in spring; male builds pebble nest. Food: Aquatic and terrestrial invertebrates, small fishes, and sometimes plants. Notes: One of the few mound-building species in headwater streams.

Cutlip Minnow

Exoglossum maxillingua

(right)

Tonguetied Minnow

Exoglossum laurae

(not shown)

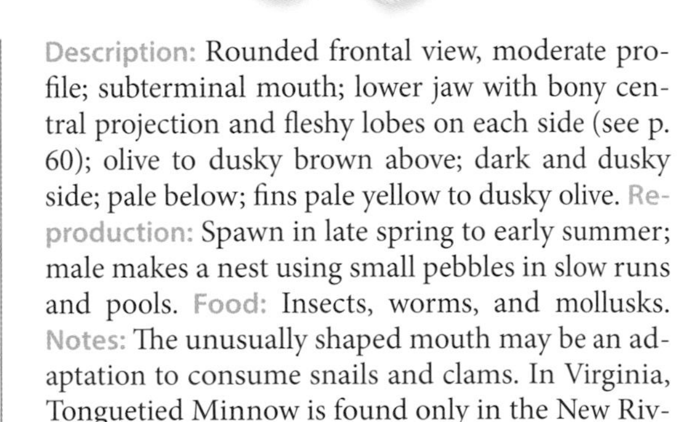

Size: 2½–6½ in.
Habitat: Cool to warm medium streams to large rivers.
Abundance: Uncommon.
Status: Native.

■ Cutlip Minnow
■ Tonguetied Minnow

Description: Rounded frontal view, moderate profile; subterminal mouth; lower jaw with bony central projection and fleshy lobes on each side (see p. 60); olive to dusky brown above; dark and dusky side; pale below; fins pale yellow to dusky olive. Reproduction: Spawn in late spring to early summer; male makes a nest using small pebbles in slow runs and pools. Food: Insects, worms, and mollusks. Notes: The unusually shaped mouth may be an adaptation to consume snails and clams. In Virginia, Tonguetied Minnow is found only in the New River system, where the species is in decline.

Bluehead Chub

Nocomis leptocephalus

Size: 3–10 in.
Habitat: Cool to warm small streams to medium rivers.
Abundance: Common.
Status: Native.

Description: Rounded frontal view, moderate profile; small barbel in corner of mouth; rounded snout; iris red; olive green to tan above, silver, brassy, yellow, or olive on side, white below; dorsal and caudal fins occasionally red; other fins yellow to white; scale margins dark; breeding male has intense blue on head; four to five large tubercles on head do not extend past nasal openings (see p. 60); orange to red lateral stripe on breeding male in all drainages except Pee Dee; female and juvenile are smaller, without tubercles. Reproduction: Spawns in spring and early summer; male constructs large pebble mound in runs and glides. Food: Aquatic and terrestrial invertebrates, small fishes, and plants. Notes: Many other minnow species (i.e., nest associates) also spawn en masse on the chub's mound, creating quite a spectacular display. Tubercles are lost after breeding, but the remaining scars can be used for identification.

River Chub

Nocomis micropogon (right)

Bigmouth Chub

Nocomis platyrhynchus
(not shown)

River Chub
Bigmouth Chub
continued:

Size: 3–12 in.
Habitat: Cool to warm small streams to large rivers.
Abundance: Common.
Status: Native.

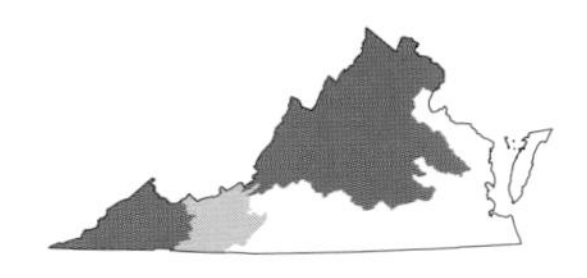

■ River Chub
■ Bigmouth Chub

Description: Rounded frontal view, moderate profile; small barbel in corner of mouth; slightly pointed snout; olive green to brown above, silver, brassy, yellow, or olive on side, white below; dorsal and caudal fins light orange to red on margins; other fins yellow to yellow olive; scale margins dark; breeding male has a pink to rosy hue over entire body, fins, and head; tubercles occur from snout to in front of eyes in River Chub and between and above eyes in Bigmouth Chub (see p. 60); females and juveniles are smaller, without tubercles. Reproduction: Spawn in spring and early summer; males construct large pebble mounds in runs and glides. Food: Aquatic and terrestrial invertebrates, small fishes, and plants. Notes: Bigmouth Chub is endemic to the New River in North Carolina, Virginia, and West Virginia. While both species appear similar, Bigmouth Chub occurs only in the New River drainage.

Bull Chub

Nocomis raneyi (not shown)

Size: 3–12 in.
Habitat: Warm small to large rivers.
Abundance: Common.
Status: Native.

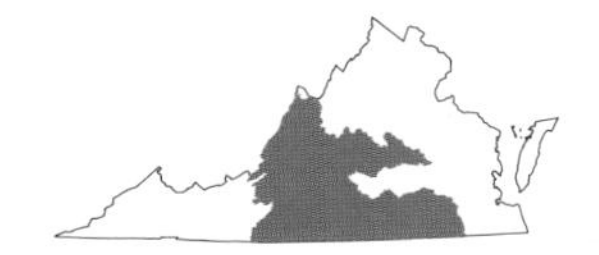

Description: Rounded frontal view, moderate profile; small barbel in corner of mouth; slightly pointed snout; mouth small compared to River and Bigmouth chubs; olive green to brown above, silver, brassy, yellow, or olive on side, white below; dorsal and caudal fins light orange to red on margins; other fins yellow to yellow olive; scale margins dark; breeding male has a pink to rosy hue over entire body, fins, and head; many small tubercles from upper lip to behind eyes (see p. 60); female and juvenile are smaller, without tubercles. Reproduction: Spawns in spring and early summer; male constructs large pebble mound in runs and glides. Food: Aquatic and terrestrial invertebrates, small fish, and plants. Notes: Bull Chub and River Chub co-occur in the James River drainage, where identification between the two can become difficult.

Streamline Chub

Erimystax dissimilis (right)

Blotched Chub

Erimystax insignis
(below right)

Slender Chub

Erimystax cahni (not shown)

Size: 3½–5¼ in.
Habitat: Warm medium streams to large rivers.
Abundance: Rare to common.
Status: Native.

■ Streamline Chub
■ Blotched Chub

Description: Slightly compressed frontal view, moderately elongate (Blotched Chub and Slender Chub) to very elongate (Streamline Chub) profile; inferior mouth; small barbel in corner of mouth; snout slightly pointed to broadly rounded; flat belly; moderate to large eye; olive above, silver on side, white to slivery white below; small spot on base of caudal fin; silver to green stripe varies on side with no blotches (Slender Chub), small round blotches within stripe (Streamline Chub), and large square blotches extending outside of stripe (Blotched Chub). **Reproduction:** Spawns in spring to early summer; eggs laid on clean silt-free gravel. **Food:** Aquatic insects, detritus, and algae. **Notes:** All three species occur only in the Tennessee drainage in Virginia. Slender Chub is extremely rare, possibly extinct, and listed as federally endangered.

Stargazing Minnow

Phenacobius uranops (right)

Fatlips Minnow

Phenacobius crassilabrum
(not shown)

Stargazing Minnow
Fatlips Minnow
continued:

Size: 2–3¾ in.
Habitat: Cold to warm medium to large rivers.
Abundance: Rare to uncommon.
Status: Native.

■ Stargazing Minnow
■ Fatlips Minnow

Description: Rounded frontal view, slightly elongate (Fatlips Minnow) to elongate (Stargazing Minnow) profile; fleshy lips similar to Suckermouth Minnow; subterminal mouth; dark stripe on back; olive to brown above, stripe on side iridescent gold to yellow to green, silver to white below; Fatlips Minnow with two yellow spots on caudal fin. Reproduction: Spawn in spring to early summer; eggs laid over gravel in riffles or on chub mounds (Fatlips Minnow). Food: Aquatic invertebrates. Notes: Members of the genus *Phenacobius* have mouths more closely resembling those of suckers than minnows. Their ranges mostly do not overlap, which simplifies their identification.

Suckermouth Minnow
Phenacobius mirabilis (right)

Kanawha Minnow
Phenacobius teretulus
(not shown)

Size: 2–3¾ in.
Habitat: Cold to warm medium to large rivers.
Abundance: Rare to uncommon.
Status: Native.

■ Suckermouth Minnow
■ Kanawha Minnow

Description: Rounded frontal view, moderate profile; fleshy lips (see p. 60); subterminal mouth; dark stripe on back; olive to brown above, stripe on side iridescent gold to yellow to green, silver to white below; Kanawha Minnow with black spots above lateral stripe. Suckermouth Minnow with rectangular black spot on base of caudal fin. Reproduction: Spawn in spring to early summer (to late summer—Suckermouth Minnow); eggs laid over gravel in riffles. Food: Aquatic invertebrates. Notes: Kanawha Minnow is endemic to the New River drainage. Suckermouth Minnow is widely distributed throughout the central United States, but the Big Sandy drainage represents the eastern-most extent of its range.

Bigeye Chub
Hybopsis amblops (right)

Highback Chub
Hybopsis hypsinotus
(not shown)

Size: 1¾–2¾ in.
Habitat: Warm small streams to medium rivers.
Abundance: Uncommon to abundant.
Status: Native.

■ Bigeye Chub
■ Highback Chub

Description: Slightly to strongly compressed frontal view, moderate (Bigeye Chub) to deep (Highback Chub) profile; short rounded snout; small inferior mouth; small barbel in corner of mouth; moderate to large upturned eye; pale olive to light yellow above, dark (Highback Chub) to light (Bigeye Chub) lateral stripe on side, white to silver below; fins generally pale; breeding males with hints of pinkish to purplish coloration. Reproduction: Spawn in late spring to early summer; Highback Chub is a nest associate of Bluehead Chub. Food: Aquatic and terrestrial invertebrates. Notes: Highback Chub is restricted to the Pee Dee drainage, whereas Bigeye Chub is present throughout the Tennessee River drainage.

Spotfin Chub
Erimonax monachus
Syn.: *Cyprinella monacha*

Size: 2½–3¾ in.
Habitat: Warm medium rivers.
Abundance: Rare.
Status: Native.
Other name: Turquoise Shiner.

Description: Compressed frontal view, elongate profile; small, inferior mouth, rounded snout that overhangs mouth; small barbel in corner of mouth; dark to faint stripe along the side; dark spot on caudal-fin base and back of dorsal fin; iridescent green stripes on back and upper side; bright silver elsewhere; breeding male is electric blue with two broad white bars or blotches on side; fins are blue with white edges; female lacks coloration.

Spotfin Chub

continued:

Reproduction: Spawns in spring and summer; eggs laid in elevated rock crevices. Food: Aquatic invertebrates. Notes: The vibrant blue coloration of the breeding male is unrivaled among US stream fishes. Spotfin Chub is listed as federally threatened and found in the Holston River system but is also historically known from the Clinch River.

Satinfin Shiner

Cyprinella analostana (right)

Spotfin Shiner

Cyprinella spiloptera

(not shown)

Steelcolor Shiner

Cyprinella whipplei

(not shown)

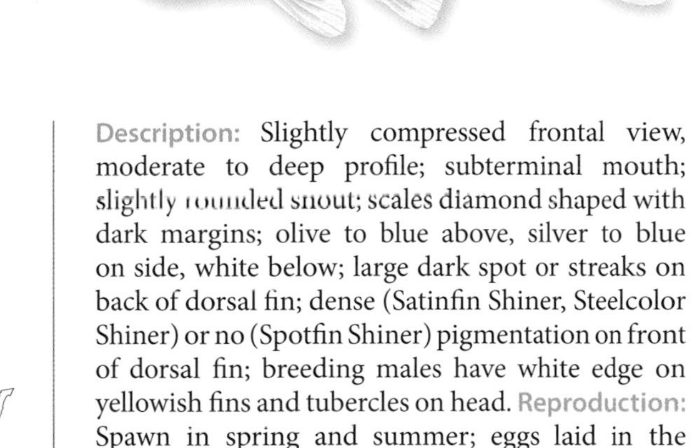

Size: 1½–4½ in.
Habitat: Warm medium streams to large rivers.
Abundance: Rare to abundant.
Status: Native.

- Satinfin Shiner
- Spotfin Shiner
- Steelcolor Shiner

Description: Slightly compressed frontal view, moderate to deep profile; subterminal mouth; slightly rounded snout; scales diamond shaped with dark margins; olive to blue above, silver to blue on side, white below; large dark spot or streaks on back of dorsal fin; dense (Satinfin Shiner, Steelcolor Shiner) or no (Spotfin Shiner) pigmentation on front of dorsal fin; breeding males have white edge on yellowish fins and tubercles on head. Reproduction: Spawn in spring and summer; eggs laid in the crevices of rocks and wood. Food: Aquatic and terrestrial invertebrates. Notes: Steelcolor Shiner is state threatened and present in the Powell and Clinch rivers. A close relative, the Thicklip Chub (*Cyprinella labrosa*), is extirpated from Virginia and occurs only in the Carolinas.

Whitetail Shiner

Cyprinella galactura

Size: 2–4 in.
Habitat: Warm medium streams to large rivers.
Abundance: Common.
Status: Native.

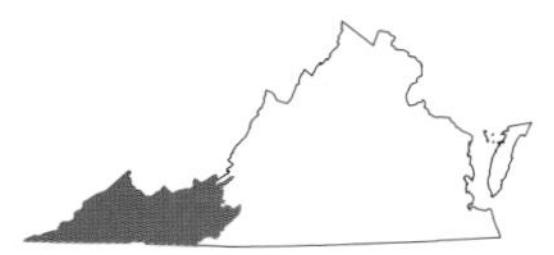

Description: Slightly compressed in frontal view, moderate to deep profile; slightly subterminal mouth; slightly rounded snout; scales diamond shaped with dark margins; olive to blue above, silver to blue on side, white below; large dark spot or streaks on back of dorsal fin; pale patches on base of caudal fin; breeding male has white edge on fins and tubercles on head. Reproduction: Spawns in spring and summer; eggs laid in the crevices of rocks and wood. Food: Aquatic and terrestrial invertebrates. Notes: Whitetail Shiner is native to the Tennessee drainage but introduced in the New River system.

Crescent Shiner

Luxilus cerasinus

Size: 2½–4 in.
Habitat: Cool to warm small streams and small rivers.
Abundance: Common.
Status: Native.

Description: Compressed frontal view, deep profile; rounded snout; scales higher than wide, especially anteriorly on body; dark stripe on back, olive to olive gray above, silver on side, silver to white below; abrupt transition from dark to light in head coloration; many irregular dark crescents on sides; breeding male with tubercles on snout with body, fins, and head becoming bright pink to magenta.

Crescent Shiner
continued:

Reproduction: Spawns in spring and early summer; nest associate on pebble nests of chubs and stonerollers. **Food:** Aquatic and terrestrial insects. **Notes:** Crescent Shiner is native to the Roanoke drainage and introduced to the James and New. The breeding male Crescent Shiner is one of the most magnificently colored fishes in Virginia.

Striped Shiner

Luxilus chrysocephalus (right)

Common Shiner

Luxilus cornutus
(not shown)

Size: 3–7 in.
Habitat: Cool to warm medium streams and large rivers.
Abundance: Common.
Status: Native.

- Striped Shiner
- Common Shiner

Description: Compressed frontal view, deep profile; rounded snout; scales higher than wide, especially forward on body; crowded nape scales (Common Shiner); dark stripes on back come together behind dorsal fin (Striped Shiner); olive above, silver on side, silver to white below; few to many dark crescents on side; breeding males with red fin margins and tubercles on head and snout; breeding males and sometimes females mostly white on side with pinkish (Common Shiner) and yellow (Striped Shiner) tones on body. **Reproduction:** Spawn in spring and early summer; nest associate on pebble nests of chubs and stonerollers. **Food:** Aquatic and terrestrial insects. **Notes:** Striped Shiner is found in the Tennessee and Big Sandy drainages and introduced in the New River.

White Shiner

Luxilus albeolus

Size: 2½–5 in.
Habitat: Cool to warm medium streams to large rivers.
Abundance: Common.
Status: Native.

Description: Compressed frontal view, moderately deep profile; rounded snout; scales higher than wide, especially forward on body; olive above, silver on side, silver to white below; zero to few dark crescents on side; breeding male blue gray above and silver on side and below, with red fins and tubercles on head and snout. Reproduction: Spawns in spring and early summer; nest associate on pebble nests of chubs and stonerollers. Food: Aquatic and terrestrial insects. Notes: White Shiner is one of the most common fish species in streams and rivers throughout montane and Piedmont regions of Virginia.

Clockwise from top: White Shiner, Central Stoneroller, Crescent Shiner, Mountain Redbelly Dace congregate in a stream.

Warpaint Shiner

Luxilus coccogenis

Size: 2¾–4 in.
Habitat: Warm small to medium rivers.
Abundance: Common.
Status: Native.

Description: Moderately compressed frontal view, elongate profile; scales higher than wide; large eye; large, terminal mouth; slightly rounded snout; olive above, upper scales with dark margins, silver on side, silver white below; white fins with black band on dorsal and caudal fins; breeding male with red-orange bar on cheek, black bar behind gill cover, red on upper lip and anterior base of dorsal fin. **Reproduction:** Spawns in spring and early summer; nest associate on chub nests. **Food:** Aquatic and terrestrial insects. **Notes:** The namesake red face markings make the Warpaint Shiner an easy-to-recognize species. Warpaint Shiner is introduced to the New River.

Mountain Shiner

Lythrurus lirus

Size: 1¾–2½ in.
Habitat: Warm large streams to medium rivers.
Abundance: Uncommon.
Status: Native.

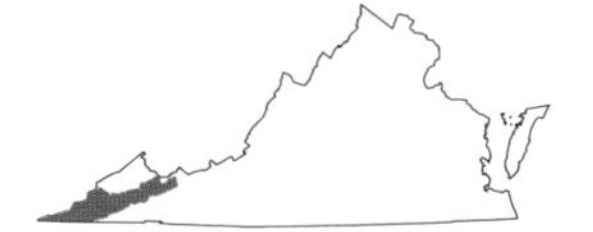

Description: Slightly compressed frontal view, elongate profile; small crowded scales behind head; snout pointed to narrowly rounded; olive above, silver on side, white below; speckling on side and back; dark silvery side stripe strong at rear and faded in front; lips black; no fin coloration. Breeding male with lemon-yellowish wash and some pinkish iridescence mid-profile. **Reproduction:** Spawns in spring to early summer; breeding habits are unknown. **Food:** Aquatic and terrestrial insects, algae, and detritus. **Notes:** Mountain Shiner is found near the water surface in large schools.

Rosefin Shiner

Lythrurus ardens (right)

Scarlet Shiner

Lythrurus fasciolaris

(not shown)

male

Size: 2–3½ in.
Habitat: Warm large streams to medium rivers.
Abundance: Uncommon to common.
Status: Native.

■ Rosefin Shiner
■ Scarlet Shiner

Description: Compressed frontal view, elongate profile; small crowded scales behind head; snout pointed to narrowly rounded; olive above, silver on side, white below; dark spot at base of dorsal fin; breeding male has tubercles on head, snout, and nape; red on all fins. Reproduction: Spawn in spring and summer; nest associate on chub and Fallfish nests. Food: Terrestrial and aquatic insects, algae, and detritus. Notes: The spot on the base of the dorsal fin is distinctive. These species are difficult to distinguish but are found in different river drainages.

Rosyface Shiner

Notropis rubellus (right)

Highland Shiner

Notropis micropteryx

(not shown)

Size: 2¼–3½ in.
Habitat: Warm medium streams to large rivers.
Abundance: Uncommon to common.
Status: Native.

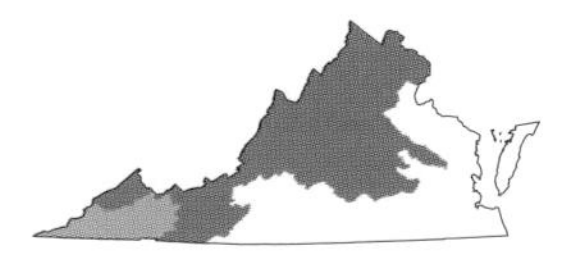

■ Rosyface Shiner
■ Highland Shiner

Description: Compressed frontal view, elongate profile; slightly pointed snout; terminal mouth; small eye; dorsal-fin origin behind origin of pelvic fin; olive gray above, silver on side, silver white below; breeding male with red orange on head, body, and base of fins. Reproduction: Spawn in spring and early summer; nest associate on chub nests. Food: Aquatic and terrestrial invertebrates. Notes: Both species can occur in large schools near the water's surface. Highland Shiner (Tennessee drainage) was recently split from Rosyface Shiner (Atlantic drainages and New River system).

Saffron Shiner

Notropis rubricroceus (right)

Redlip Shiner

Notropis chiliticus
(not shown)

Size: 1¾–3 in.
Habitat: Cold to warm small streams to large rivers.
Abundance: Uncommon to abundant.
Status: Native.

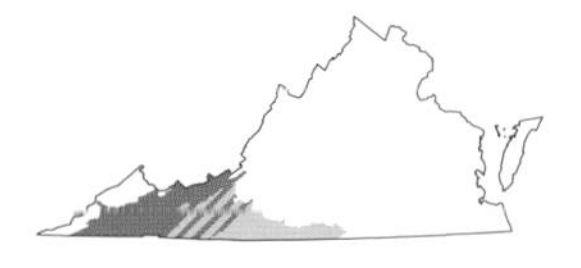

■ Saffron Shiner
■ Redlip Shiner

Description: Slightly compressed frontal view, moderate profile; rounded snout; slightly subterminal mouth; dark, posterior stripe; dark rectangular spot at base of caudal fin; side with darkly outlined scales (Saffron Shiner) or scattered black blotches (Redlip Shiner); olive above, silver on side, silver white below; fins pale; breeding males of Redlip Shiner and Saffron Shiner have bright yellow fins; Redlip Shiner with iridescent gold to yellow stripe on side and bright red lips, scarlet body; Saffron Shiner with dark red above to scarlet below. Reproduction: Spawn in spring and early summer; nest associate on chub and stoneroller nests. Food: Aquatic and terrestrial invertebrates. Notes: Redlip Shiner is native to the Pee Dee drainage but has been introduced into the New River system. Saffron Shiner is native to the Tennessee drainage and has been introduced into the New River. During the spring, both species are colorful residents of Virginia's mountain streams.

Tennessee Shiners in breeding coloration.

Tennessee Shiner

Notropis leuciodus

Size: 1¾–3 in.
Habitat: Cool to warm small streams and large rivers.
Abundance: Abundant.
Status: Native.

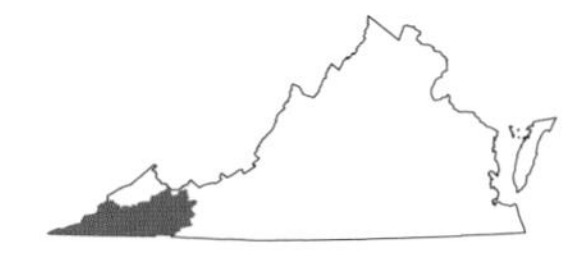

Description: Slightly compressed frontal view, elongate profile; snout slightly pointed; terminal mouth; dark posterior stripe; dark rectangular spot at base of caudal fin; darkly stitched lateral line; olive above, silver on side, silver white below; fins pale; breeding male is bright orange red. **Reproduction:** Spawns in spring and early summer; nest associate on chub and stoneroller nests. **Food:** Aquatic and terrestrial invertebrates. **Notes:** Tennessee Shiner is introduced into the New River drainage but is rarely found. The color change during the breeding season from a dull silvery minnow to a bright red minnow is incredible.

Silver Shiner

Notropis photogenis (right)

Comely Shiner

Notropis amoenus (not shown)

Emerald Shiner

Notropis atherinoides (not shown)

Size: 2½–5½ in.
Habitat: Warm medium to large rivers.
Abundance: Rare to uncommon.
Status: Native.

- Silver Shiner
- Comely Shiner
- Emerald Shiner

Description: Compressed frontal view, elongate profile; pointed snout; terminal mouth; moderate-sized eye; dorsal-fin origin behind that of pelvic fins; olive to gray to green above, side silver with brassy (Comely Shiner), gold (Silver Shiner), or greenish (Emerald Shiner) stripe, silver below; Silver Shiner has black lips and two dark crescents between nostrils; fins pale. **Reproduction:** Spawn in spring and summer; breeding is thought to occur at night. **Food:** Aquatic and terrestrial invertebrates.

Silver Shiner
Comely Shiner
Emerald Shiner
continued:

Notes: Silver Shiner is common in the New and Tennessee drainages and rare in the Big Sandy. Although occurring widely in North America, Emerald Shiner is extremely rare in Virginia, with only a few records in the Clinch and Powell rivers. Comely Shiner occurs somewhat uncommonly in Atlantic drainages.

Telescope Shiner
Notropis telescopus

Size: 2–2½ in.
Habitat: Warm medium streams to large rivers.
Abundance: Abundant.
Status: Native.

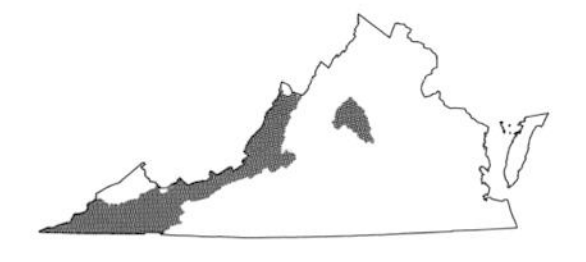

Description: Slightly compressed frontal view; elongate profile; large, round eye; zig-zag pattern between scales on back near head; snout slightly rounded; mouth terminal; light to dark olive above, silver on side, silver white below; darkly stitched lateral line. Reproduction: Spawns in spring and summer; spawning habits require further study. Food: Immature aquatic insects and other invertebrates. Notes: Telescope Shiner was introduced into the New and James rivers, most likely through bait bucket releases.

Popeye Shiner
Notropis ariommus (right)

Roughhead Shiner
Notropis semperasper (not shown)

Sandbar Shiner
Notropis scepticus (not shown)

(*continued*)

Popeye Shiner
Roughhead Shiner
Sandbar Shiner
continued:

Size: 2–2½ in.
Habitat: Warm small to large rivers.
Abundance: Rare to abundant.
Status: Native.

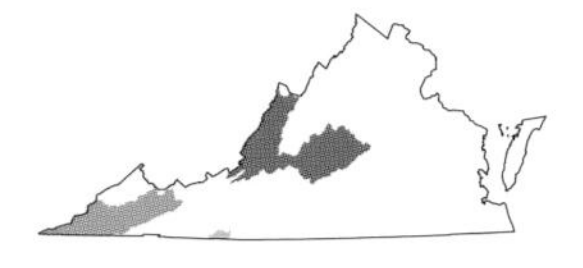

- Popeye Shiner
- Roughhead Shiner
- Sandbar Shiner

Description: Slightly compressed frontal view; elongate (Roughhead Shiner); relatively deep (Popeye Shiner adults) to deep (Sandbar Shiner) profile; large round eye; snout slightly rounded (Roughhead Shiner) to moderately pointed (Popeye Shiner and Sandbar Shiner); mouth terminal (Popeye Shiner and Sandbar Shiner) or slightly subterminal (Roughhead Shiner); Roughhead Shiner dorsal-fin origin slightly behind pelvic-fin origin; light to dark olive above, silver on side, silver white below. Reproduction: Spawn in spring and summer; spawning habits require further study. Food: Immature aquatic insects and other invertebrates. Notes: Popeye Shiner is a species of very high conservation need in Virginia. Roughhead Shiner is endemic and becoming rare in the upper James drainage, possibly due to competition with introduced Telescope Shiner. Sandbar Shiner was only recently discovered in Virginia's section of the Pee Dee drainage.

Spottail Shiner
Notropis hudsonius

Size: 2½–5¾ in.
Habitat: Warm medium streams to large rivers.
Abundance: Common.
Status: Native.

Description: Slightly compressed frontal view; moderate profile; rounded snout; subterminal mouth; large slightly upturned eye; olive above, side silver, silver white below; dark spot at base of caudal fin. Reproduction: Spawns in spring; large groups deposit eggs over sand and gravel substrate. Food: Aquatic and terrestrial invertebrates, small fish, and plant matter. Notes: Spottail Shiner occurs throughout the Atlantic slope drainages and was introduced into the New River. It is a schooling species that provides a food source for predatory fishes.

Mimic Shiner

Notropis volucellus (right)

New River Shiner

Notropis scabriceps
(not shown)

Size: 1½–3¾ in.
Habitat: Warm medium streams to large rivers.
Abundance: Rare to uncommon.
Status: Native.

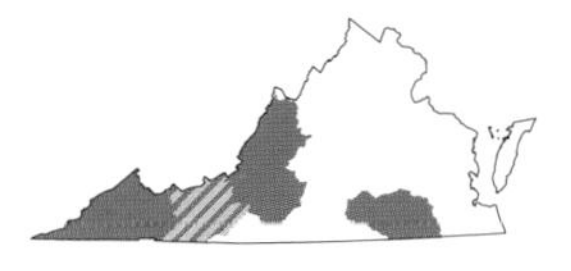

■ Mimic Shiner
■ New River Shiner

Description: Slightly compressed frontal view; moderate profile; subterminal mouth; broad snout; eye is large and upturned (New River Shiner) or smaller and lateral (Mimic Shiner); stripe on side is darkest at rear; lateral-line scale height anteriorly noticeably tall relative to width (Mimic Shiner) or normal (New River Shiner); light yellow olive above, silver on side, silver white below; fins pale. Reproduction: Spawn in spring and summer; likely spawn multiple times per year. Food: Aquatic and terrestrial invertebrates, plants, and detritus. Notes: New River Shiner is endemic to the New River drainage. As the name implies, Mimic Shiners appear similar and may be easily confused with other species, especially the New River, Mirror, Sand, and Sawfin shiners.

Sawfin Shiner

Notropis sp. cf. *spectrunculus*
(right)

Mirror Shiner

Notropis spectrunculus
(not shown)

Size: 1¾–2¾ in.
Habitat: Cold to warm medium streams and rivers.
Abundance: Rare to common.
Status: Native.

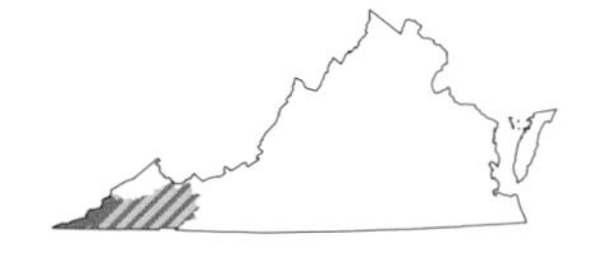

■ Sawfin Shiner
■ Mirror Shiner

Description: Rounded frontal view, moderate profile; subterminal mouth; pectoral-fin rays with rows of tubercles (breeding male Sawfin); dark stripe posteriorly; olive above, silver on side, silver white below; spot at base of caudal fin dark (Mirror) to faded (Sawfin); lacks pigment on caudal fin (Sawfin), or all fins pigmented (Mirror); pectoral, dorsal, pelvic, and anal fins in breeding male orange to bright red (Sawfin) to dark and dusky with orange pink in dorsal and caudal fins (Mirror).

Sawfin Shiner
Mirror Shiner
continued:

Reproduction: Spawn in late spring and early summer; spawning habits require further study. Food: Aquatic invertebrates and detritus. Notes: Mirror Shiner is sporadically distributed throughout tributaries of the upper Clinch and Holston rivers. Sawfin Shiner is found throughout the Tennessee River drainage.

Swallowtail Shiner

Notropis procne (right)

Sand Shiner

Notropis stramineus
(not shown)

Whitemouth Shiner

Notropis alborus
(not shown)

Size: 1½–2¾ in.
Habitat: Warm small streams to large rivers.
Abundance: Common to uncommon.
Status: Native.

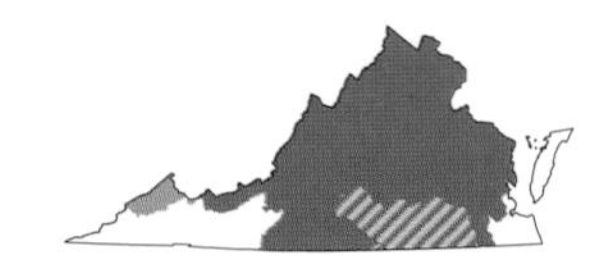

- Swallowtail Shiner
- Sand Shiner
- Whitemouth Shiner

Description: Slight to moderately compressed frontal view, elongate to moderate profile; moderate to large eye; snout rounded (Whitemouth Shiner) to slightly rounded (Swallowtail Shiner) to slightly pointed (Sand Shiner); subterminal mouth; dark stripe beginning at snout and extending to caudal fin; Whitemouth Shiner with a dark band of pigmentation encircling side and front of snout but not touching upper lip (see p. 61); Swallowtail Shiner with dark pigmentation on side but lighter in front of snout (see p. 61); Sand Shiner evenly pigmented on side and front of snout; spot on base of caudal fin; straw to olive above, straw on side, white to silver below. Reproduction: Spawn from early spring to late summer; eggs are laid in sand and gravel in moderate current. Food: Aquatic and terrestrial invertebrates. Notes: Swallowtail Shiner is introduced in the New River. Whitemouth Shiner is listed as state threatened. Sand Shiner can be easily confused with Mimic Shiner where the two co-occur.

Bridle Shiner

Notropis bifrenatus

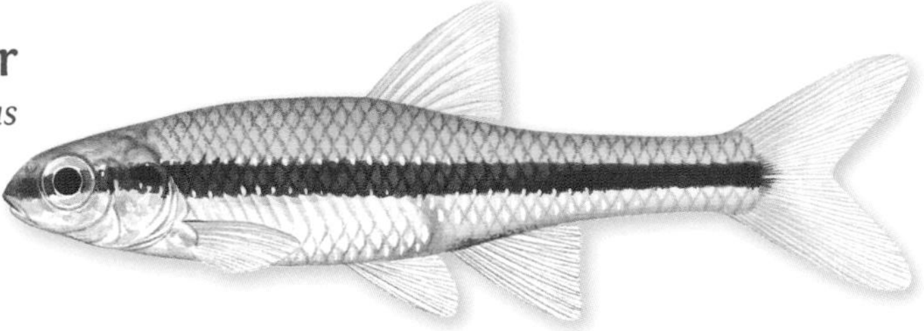

Size: 1½–2¾ in.
Habitat: Warm small streams to large rivers.
Abundance: Rare.
Status: Native.

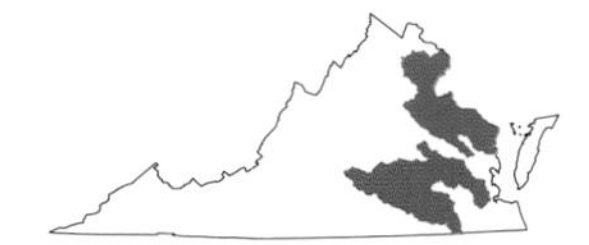

Description: Moderately compressed frontal view, moderate profile; large eye; slightly rounded snout; subterminal mouth; prominent stripe beginning at snout and extending to caudal fin; front of upper lip well pigmented (see p. 61); spot on base of caudal fin; straw to olive above, straw on side, white to silver below. **Reproduction:** Spawns from early spring to late summer; preferred mating habitat is sluggish pools with aquatic vegetation. **Food:** Aquatic and terrestrial invertebrates. **Notes:** Bridle Shiner has become exceedingly rare in Virginia and throughout its range.

Ironcolor Shiner

Notropis chalybaeus (right)

Highfin Shiner

Notropis altipinnis
(not shown)

Size: 1½–2¾ in.
Habitat: Warm small streams to large rivers and swamps (Ironcolor Shiner).
Abundance: Uncommon.
Status: Native.

- Ironcolor Shiner
- Highfin Shiner

Description: Slight to moderately compressed frontal view, elongate to moderate profile; large eye; slightly rounded (Highfin Shiner) to slightly pointed (Ironcolor Shiner) snout; terminal mouth; dorsal-fin origin above (Ironcolor Shiner) to slightly behind (Highfin Shiner) pelvic-fin origin; lower lip, chin, and snout well pigmented (see p. 61); dark stripe beginning at snout and extending to caudal fin; spot on base of caudal fin; straw to olive above, straw on side, white to silver below.

(continued)

Ironcolor Shiner
Highfin Shiner
continued:

Reproduction: Spawn from early spring to late summer; preferred mating habitat is sluggish, sandy pools. Food: Aquatic and terrestrial invertebrates. Notes: These species are unique among minnows because they can tolerate acidic conditions of the eastern Piedmont and Coastal Plain.

Silverjaw Minnow

Notropis buccatus
Syn.: *Ericymba buccata*

Size: 1½–3 in.
Habitat: Warm medium streams to large rivers.
Abundance: Uncommon.
Status: Native.

Description: Slightly compressed frontal view, elongate profile; several large silver chambers on cheek and lower jaw; head flat on bottom; subterminal mouth; snout slightly rounded; large upturned eye; tan to light olive above, silver on side, white below; fins clear. Reproduction: Spawns in spring and summer; mates in groups on gravel-sand bottom. Food: Aquatic insects, mollusks, crustaceans, detritus, and algae. Notes: The distinctive chambers on the head of the Silverjaw Minnow may be used to detect prey movement.

Eastern Silvery Minnow

Hybognathus regius

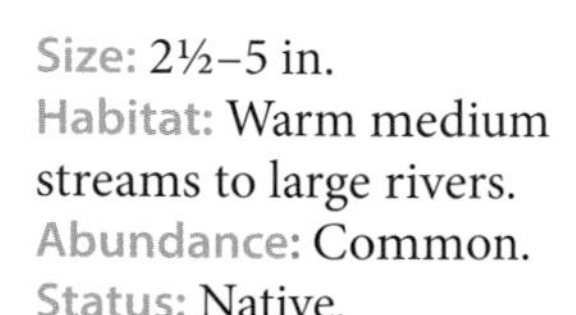

Size: 2½–5 in.
Habitat: Warm medium streams to large rivers.
Abundance: Common.
Status: Native.

Description: Compressed frontal view, moderate profile; small slightly subterminal mouth downcurved posteriorly; thick caudal peduncle; yellow olive to brown above, silver on side, brassy to silver below; light-brown fins. Reproduction: Spawns in spring; spawns communally on unvegetated areas of silt, sand, or gravel.

Eastern Silvery Minnow *continued:*

Food: Algae, diatoms, and detritus. Notes: Eastern Silvery Minnow may be found in fresh and slightly brackish waters. The species typically occurs in large schools, especially during spawning.

Bluntnose Minnow

Pimephales notatus (right)

Bullhead Minnow

Pimephales vigilax (not shown)

Fathead Minnow

Pimephales promelas (not shown)

male

Size: 1½–4 in.
Habitat: Warm medium streams to large rivers and ponds.
Abundance: Rare to abundant.
Status: Native.

■ Bluntnose Minnow
■ Bullhead Minnow

Description: Rounded frontal view, moderate (Fathead Minnow) to slightly elongate profile (Bluntnose Minnow and Bullhead Minnow); terminal (Bullhead Minnow and Fathead Minnow) or subterminal (Bluntnose Minnow) mouth; scales crowded behind head; short blunt snout; spot on front of dorsal fin; incomplete (Fathead Minnow) or complete (Bluntnose Minnow and Bullhead Minnow) lateral line; olive to tan above; dark to dusky stripe on side; silver white below; breeding males with tubercles on snout; head black; dark bands on dorsal and caudal fins and purplish sheen overall on Fathead Minnow. Reproduction: Spawn in spring, summer, and possibly fall; males build a nest under rock or wood. Food: Aquatic invertebrates, plants, and detritus. Notes: Fathead Minnow (sometimes sold as "toughies") is introduced throughout Virginia from bait buckets and may be common in ponds. Bullhead Minnow is rare and known only from the lower portions of the Tennessee drainage.

SUCKERS
Family Catostomidae

Suckers, family Catostomidae, possess a single soft-rayed dorsal fin, a sub-terminal mouth, soft fleshy lips, and large paired fins. The common name, sucker, refers to the downward-directed mouth that serves to suck up small benthic organisms. The Commonwealth is known for 19 living and 1 extinct (Harelip Redhorse, *Moxostoma lacerum*) sucker species. The suckers of Virginia range from deep-bodied Quillback that lives in large rivers to the small Torrent Sucker that occupies riffles in small headwaters. Males often possess tubercles on their anal and lower caudal fins during breeding. Suckers are generally found in schools and migrate upstream prior to breeding. Although they have a reputation for being bottom feeders and egg eaters, their lips contain numerous taste buds for selective feeding, and they consume a wide variety of small insects, algae, mollusks, and detritus. Any unwanted food material is sensed by the palatal organ in the throat and ejected through the gills. The presence and numbers of suckers is one indicator of good water quality in streams and rivers. Some suckers in Virginia remain small through maturity, while others, such as the redhorses, can attain lengths of up to 30 inches. All suckers have good tasting, though bony, flesh.

Lip forms of suckers:

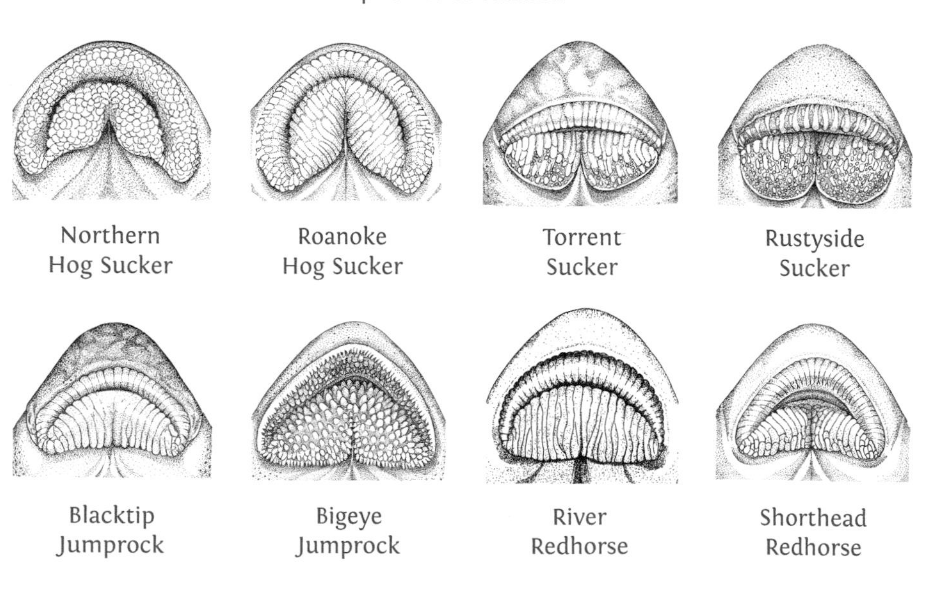

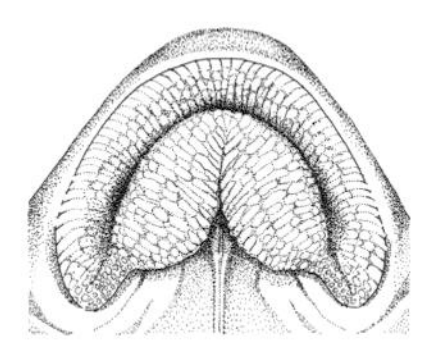

Notchlip and Silver Redhorse

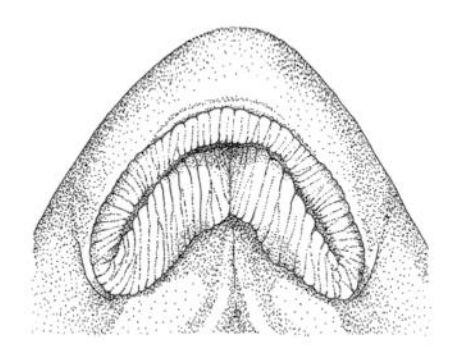

Golden Redhorse

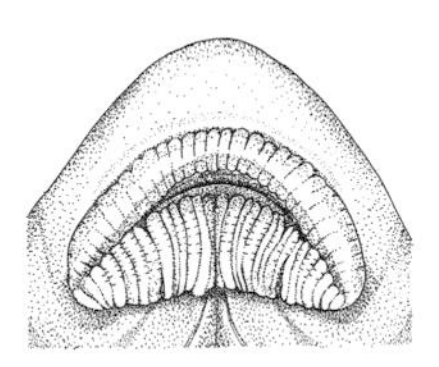

Black Redhorse

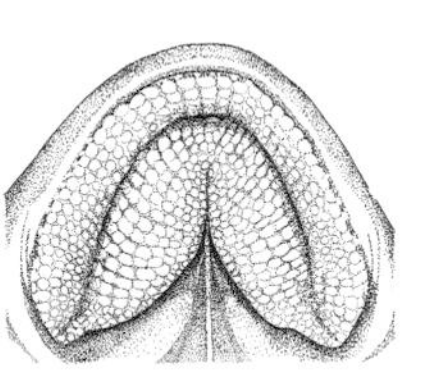

V-lip Redhorse

Quillback

Carpiodes cyprinus

Size: 8–16 in.
Habitat: Warm large rivers and reservoirs.
Abundance: Uncommon.
Status: Native.

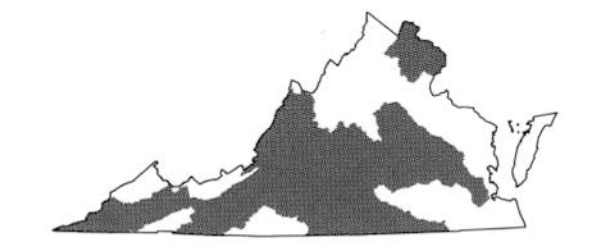

Description: Compressed frontal view, deep profile; highbacked; conical head; nipple absent in the middle of the lower lip; large scales; long first dorsal fin extended almost to caudal fin; olive to brown above, silver and golden on sides, white below, fins gray with white edges. **Reproduction:** Spawns in spring; migrates into tributaries to spawn over gravel, sand, or mud. **Food:** Small mollusks, aquatic invertebrates, detritus, and algae. **Notes:** Quillback is reported to be a strong fighter for those lucky enough to catch it on a hook and line.

White Sucker

Catostomus commersonii

Size: 10–20 in.
Habitat: Cool to warm small streams to large rivers, ponds, reservoirs.
Abundance: Abundant.
Status: Native.

Description: Rounded frontal view, moderate profile; deep median notch in the middle of the lower lip; papillose lips; scales on side are crowded together near the operculum and are larger toward caudal fin; olive brown above with white gray underside. Reproduction: Spawns in spring; migrates into tributaries to spawn over clean gravel. Food: Small crustaceans, aquatic invertebrates, detritus, algae. Notes: White Sucker is one of Virginia's most abundant and widespread sucker species. Anglers use dip nets to capture White Sucker during spring migration.

Eastern Creek Chubsucker

Erimyzon oblongus (right)

Lake Chubsucker

Erimyzon sucetta
(not shown)

Size: 10–20 in.
Habitat: Small streams and rivers, swamps, ponds, reservoirs (Creek); warm sluggish streams, swamps, ponds, reservoirs (Lake).
Abundance: Common to uncommon.
Status: Native.

■ Eastern Creek Chubsucker
■ Lake Chubsucker

Description: Moderately compressed frontal view, moderate to deep profile; adult males can develop thorny snout tubercles; plicate lips, which meet at nearly right angles; large scales and no lateral line; distinct dark lateral stripe in juveniles; color varies from brown to olive green above and silver or white below. Reproduction: Spawn in spring; scatter spawn over vegetation or on clean gravel substrate. Food: Planktonic crustaceans, small insects. Notes: Chubsuckers lack a lateral line, a sensory organ found in most fish species. The young may be confused with some cyprinids due to lateral stripe.

Northern Hog Sucker

Hypentelium nigricans (right)

Roanoke Hog Sucker

Hypentelium roanokense (not shown)

Size: 5–16 in.
Habitat: Cool to warm small to medium rivers.
Abundance: Uncommon to common.
Status: Native.

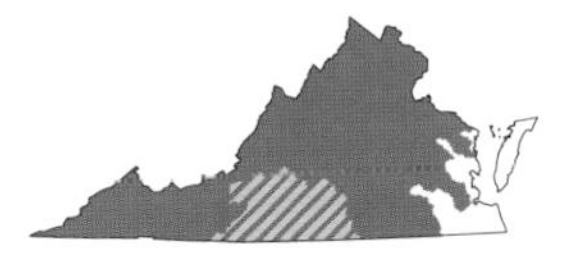

■ Northern Hog Sucker
■ Roanoke Hog Sucker

Description: Rounded frontal view, elongate profile; Northern Hog Sucker has a blocky head, concave area between eyes, dark saddles, papillose lips; Roanoke Hog Sucker rarely exceeds five inches in length, are stockier with a freckled snout, and lips are papillose on outer surface and plicate on inner surface (see p. 90); bronze dorsally and white ventrally; ventral fins olive to light orange. Reproduction: Spawn in spring; highly migratory; a single female can attract three to six males and maintain spawning position at the end of pools. Food: Microcrustaceans, immature aquatic insects, and algae. Notes: Hog Suckers use their large pectoral fins to provide stabilization in swift current. Roanoke Hog Sucker is endemic to the Roanoke River but has recently been discovered in the Pee Dee drainage.

Torrent Sucker

Thoburnia rhothoeca (right)

Rustyside Sucker

Thoburnia hamiltoni (not shown)

Size: 3–6 in.
Habitat: Cold to warm small streams to small rivers.
Abundance: Uncommon to abundant.
Status: Native.

Description: Rounded frontal view, elongate profile; plicate-papillose lips; the lobes of the lower lip are round in Rustyside Sucker and angular in Torrent Sucker (see p. 90); two pale spots at base of caudal fin; black blotches on sides; breeding males have brilliant rusty-orange stripes down each side.

(*continued*)

Torrent Sucker
Rustyside Sucker
continued:

■ Torrent Sucker
■ Rustyside Sucker

Rustyside Sucker has smaller eyes and a wider, bumpier lower lip than Torrent Sucker. **Reproduction:** Spawn in spring; most likely lay eggs in riffles. **Food:** Algae, detritus, and small aquatic invertebrates. **Notes:** Rustyside Sucker is known only from the upper Dan River in the Roanoke River drainage. Both species are well adapted to holding their position in swift water due to large paired fins, a diminutive gas bladder, and streamlined body.

Blacktip Jumprock
Moxostoma cervinum (right)

Brassy Jumprock
Moxostoma sp. cf. *lachneri*
(not shown)

Size: 4–8 in.
Habitat: Warm small to medium rivers.
Abundance: Uncommon to abundant.
Status: Native.

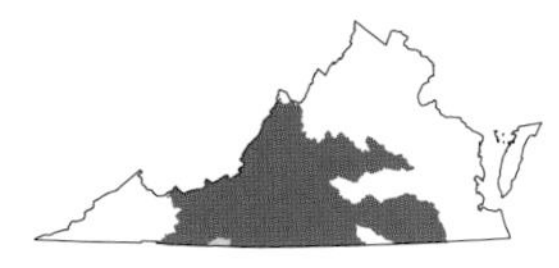

■ Blacktip Jumprock
■ Brassy Jumprock

Description: Rounded frontal view, elongate profile; plicate lips (see p. 90); gray olive above, brassy, yellow, silver iridescence, and violet hues on sides (Blacktip Jumprock), brassy, olive silver, and green-golden iridescence on sides (Brassy Jumprock), white pearly below; light and dark stripes on back and sides (evident only in juvenile of Brassy Jumprock); black pigmentation on distal edges of dorsal and caudal fins (Blacktip Jumprock); breeding males exhibit small tubercles on anal and lower caudal-fin margins. **Reproduction:** Spawn in late winter to late spring; eggs are laid in swift water. **Food:** Immature aquatic insects, especially midge larvae. **Notes:** Brassy Jumprock is larger than Blacktip Jumprock. Brassy Jumprock was previously known as Smallfin Redhorse. Blacktip Jumprock was introduced via bait buckets to the New River.

Bigeye Jumprock

Moxostoma ariommum

Size: 4–8 in.
Habitat: Cool to warm medium rivers.
Abundance: Uncommon.
Status: Native.

Description: Rounded frontal view, elongate profile; papillose lips (see p. 90); four lateral blotches; gray olive above, green gray on sides, white pearly below; large eyes; breeding male exhibits small tubercles on anal and lower caudal-fin margins. Reproduction: Spawns in late winter to late spring; likely lays eggs in swift water. Food: Immature aquatic insects and detritus. Notes: Large eyes may be an adaptation for defense against avian predators. Bigeye Jumprock collected under banks or woody debris tend to be darker in color.

River Redhorse

Moxostoma carinatum

Size: 18–26 in.
Habitat: Warm medium to large streams to large rivers.
Abundance: Common.
Status: Native.

Description: Rounded frontal view, moderate profile; head large; lips deeply plicate (see p. 90); snout blunt; olive to tan above, yellow on sides, silver flank white below; dorsal fin slightly concave; caudal fin deep red, paired fins orange; fin color more intense in adult. Reproduction: Spawns in early spring to early summer; spawns in beds of gravel and rubble in shallow riffles and runs. Food: Insects, crustaceans, mollusks, and detritus. Notes: River Redhorse has a pointed upper caudal lobe; the lower lobe is rounded. Internal teeth are adapted to crush shellfish and crayfish. It is considered excellent table fare.

Shorthead Redhorse

Moxostoma macrolepidotum (right)

Smallmouth Redhorse

Moxostoma breviceps
(not shown)

Size: 12–18 in.
Habitat: Warm medium to large streams to large rivers.
Abundance: Common.
Status: Native.

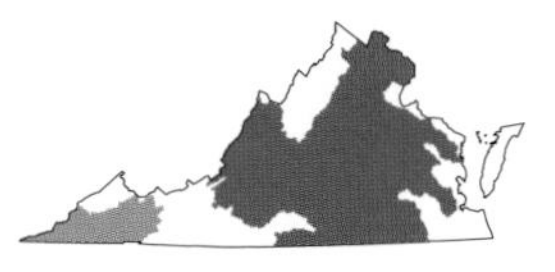

■ Shorthead Redhorse
■ Smallmouth Redhorse

Description: Rounded frontal view, moderate profile; high back; short head; small mouth and lips; lips plicate (see p. 90); snout blunt; dorsal fin slightly (Shorthead Redhorse) to deeply (Smallmouth Redhorse) concave; dark olive to tan above, silver, yellow, golden on sides, scale bases dark, white below; dorsal and caudal fins deep red, paired fins pale red to pale orange; upper caudal lobe longer than lower in Smallmouth Redhorse; fin color more intense in adults. **Reproduction:** Spawn in early spring to early summer; breed in beds of gravel and cobble in shallow riffles and runs. **Food:** Insects, crustaceans, mollusks, detritus, and algae. **Notes:** Shorthead Redhorse is widespread in Atlantic Slope rivers from New York to South Carolina. Smallmouth Redhorse is found throughout the Mississippi drainage.

Silver Redhorse

Moxostoma anisurum (right)

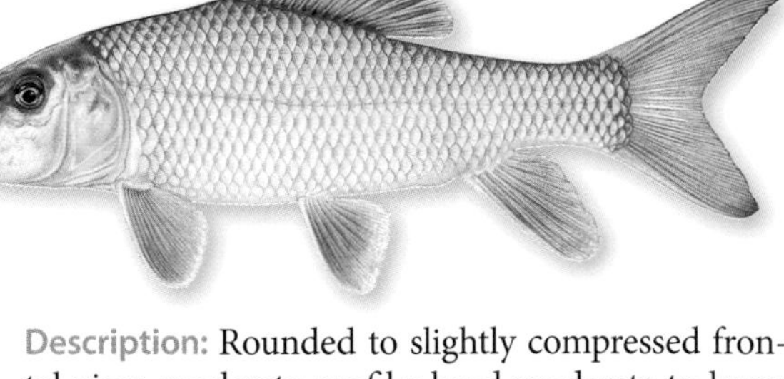

Notchlip Redhorse

Moxostoma collapsum
(not shown)

Size: 18–24 in.
Habitat: Warm medium streams to large rivers, reservoirs.
Abundance: Rare to uncommon.
Status: Native.

Description: Rounded to slightly compressed frontal view, moderate profile; head moderate to large; plicate-papillose lips, V-shaped lower lip (see p. 91); slightly convex dorsal-fin edge; Silver Redhorse has a pointed upper caudal lobe, while the lower is rounded; olive or tan above, silver, yellow, gold, brassy on sides, white below; dorsal and caudal fins olive or dusky; paired fins orange to pale red.

Silver Redhorse
Notchlip Redhorse
continued:

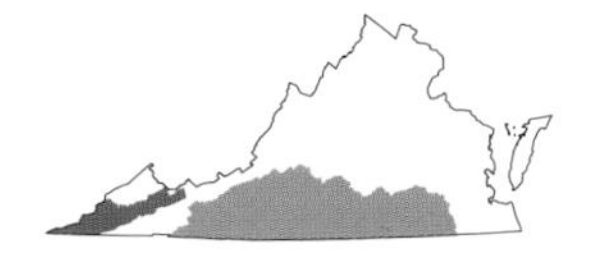

■ Silver Redhorse
■ Notchlip Redhorse

Reproduction: Spawn in early spring; breed in beds of gravel and rubble in shallow riffles and runs. **Food:** Aquatic invertebrates, mollusks, algae, and detritus. **Notes:** Notchlip Redhorse has recently been discovered in the New River drainage, a result of introduction. Both species are intolerant to pollution. Notchlip Redhorse may be confused with the V-lip Redhorse, which has more rounded papillae on lips and concave dorsal fin.

Golden Redhorse

Moxostoma erythrurum

Size: 12–18 in.
Habitat: Warm medium streams to large rivers, reservoirs.
Abundance: Common.
Status: Native.

Description: Round to slightly compressed frontal view, moderate profile; head moderate size; lips plicate, lower lip slightly V shaped (see p. 91); olive tan above, silver, yellow, gold, brassy on sides, white below; dorsal-fin margin slightly concave, caudal fin dusky dark with pointed equal lobes, paired fins orange to pale red. **Reproduction:** Spawns in early spring to early summer; breeds in beds of gravel and cobble in shallow riffles and runs. **Food:** Aquatic invertebrates, algae, and detritus. **Notes:** Golden Redhorse is one of the most widely distributed of the redhorses in Virginia. It is introduced into the Potomac drainage.

Black Redhorse

Moxostoma duquesnei

Size: 10–15 in.
Habitat: Cool to warm medium streams to large rivers, reservoirs.
Abundance: Common.
Status: Native.

Description: Rounded to slightly compressed frontal view, elongate profile; head long; lips plicate, lower lip straight (see p. 91); olive tan above, silver, gold, or pale green iridescence on sides, white below; dorsal edge slightly concave, dorsal and caudal fins dark, caudal lobes are equal and pointed, paired fins orange to pale red. **Reproduction:** Spawns in late spring to early summer; breeds in beds of gravel and rubble in shallow riffles and runs. **Food:** Aquatic invertebrates, algae, and detritus. **Notes:** The species name originates from Fort Duquesne, now known as Pittsburgh, Pennsylvania. The identification between Black Redhorse and Golden Redhorse can be difficult where they co-occur.

V-lip Redhorse

Moxostoma pappillosum

Size: 9–13 in.
Habitat: Warm medium streams to medium rivers.
Abundance: Uncommon.
Status: Native.

Description: Rounded to slightly compressed frontal view, elongate profile; head long; lower lip deep V shape with round papillae (see p. 91); olive tan above, silver, gold, brassy on sides, white below; dorsal edge concave, dorsal and caudal fins dark, paired fins orange to pale red. **Reproduction:** Spawns in late spring to early summer; breeds in beds of gravel and rubble in shallow riffles and runs.

V-lip Redhorse
continued:

Food: Aquatic invertebrates, algae, and detritus. **Notes:** V-lip Redhorse uses pectoral fins to flip rocks in search of food. It is most abundant in the upper Roanoke drainage.

An aggregation of Golden Redhorse during spawning.

NORTH AMERICAN CATFISHES
Family Ictaluridae

Endemic to North America, the family Ictaluridae is composed of 51 species of catfishes, 15 of which inhabit Virginia's waters. The most recognizable family trait is the four pairs of barbels (or whiskers) that are used to taste their surroundings. Other characteristics include a scaleless body, single spines on the dorsal and pectoral fins, and presence of an adipose fin. Some species exceed lengths of 3 feet and weights of 100 pounds, and most are good table fare, making them popular for recreational fisheries. In contrast, almost half of Virginia's catfish species are known as madtoms that grow to only a few inches. Although small, the nongame madtoms have needle sharp, mildly venomous spines on their dorsal and pectoral fins. The success of catfishes is related to their adaptability to a wide variety of habitats, prey bases, and water quality conditions. However, some species, in particular madtoms, have specific habitat requirements and are especially sensitive to pollution and habitat degradation.

Adipose-fin shapes of North American Catfishes:

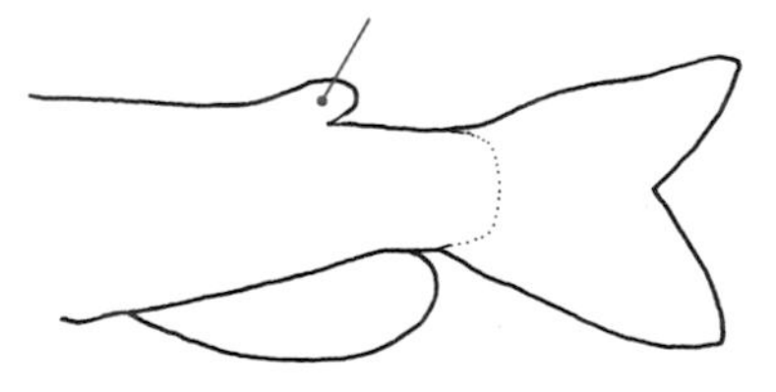

Catfishes and bullheads: separate

Madtoms: connected

Tail-fin shapes of North American Catfishes:

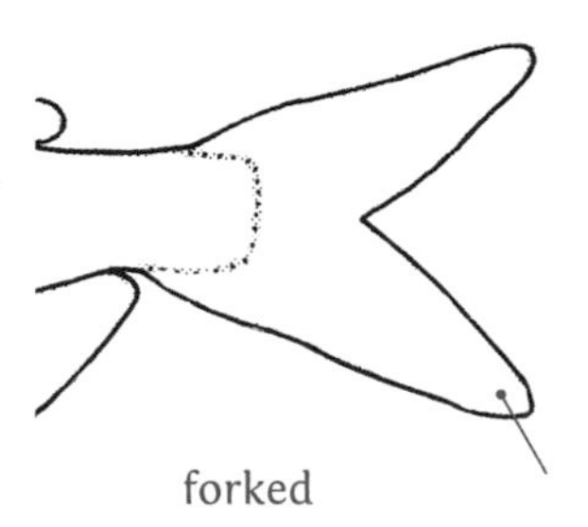

forked

emarginate

Blue Catfish

Ictalurus furcatus

Size: 12 in. to 4 ft.
Habitat: Large rivers, reservoirs, and estuaries.
Abundance: Uncommon to extremely abundant.
Status: Introduced.

Description: Round frontal view, compressed toward tail; moderate profile; wedge-shaped head; subterminal or slightly inferior mouth; deeply forked tail; adipose fin separate from caudal fin (see p. 100); straight-edged anal fin, usually with 30–36 rays; blue or grey above, silver to white on sides without dark spots, white below. **Reproduction:** Spawns in late spring and early summer; spawning occurs in cavities guarded by males. **Food:** Aquatic plants and invertebrates including mollusks and fishes. **Notes:** Blue Catfish was introduced to the Atlantic Slope drainages, and several Virginia rivers support large populations. The world record for Blue Catfish (143 lb.) was caught at Kerr Reservoir in 2011.

Channel Catfish

Ictalurus punctatus

Size: 11 in. to 2.5 ft.
Habitat: Large rivers, warm streams, ponds, lakes, and reservoirs, including estuarine habitats.
Abundance: Uncommon to common.
Status: Native.

Description: Round frontal view, compressed toward tail; moderate profile; depressed head; subterminal or inferior mouth; deeply forked caudal fin (see p. 100); adipose fin separate from caudal fin; rounded anal fin, usually with 25–30 rays; blue, gray, or olive-green above, slate blue to gray on sides, sometimes with dark spots, yellow or white below.

(*continued*)

Channel Catfish
continued:

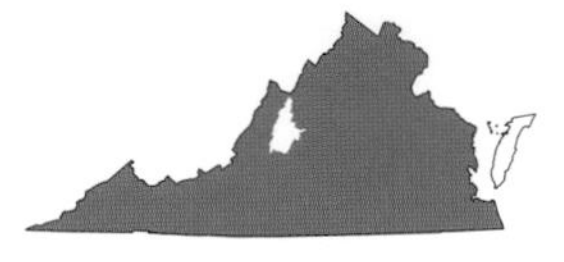

Reproduction: Spawns in late spring and early summer; spawning occurs in cavities guarded by males. Food: Aquatic invertebrates, fishes, and plants. Notes: Channel Catfish is widely distributed across Virginia, but likely only native to the Big Sandy, New, and Tennessee river drainages. Channel Catfish has been widely introduced for its sporting and culinary quality.

White Catfish
Ameiurus catus

Size: 8–18 in.
Habitat: Ponds, reservoirs, warm rivers, and estuaries.
Abundance: Uncommon to common.
Status: Native.

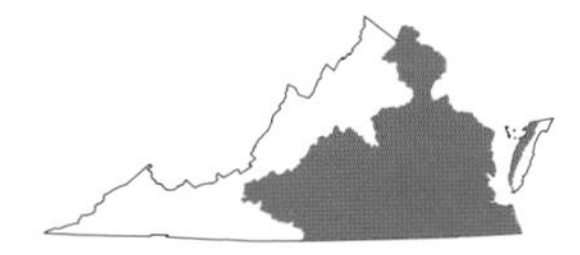

Description: Round frontal view, compressed toward tail; moderate to elongate profile; large, depressed head; subterminal or slightly inferior mouth; moderately forked tail; adipose fin separate from caudal fin (see p. 100); rounded anal fin; blue or gray above, fading to white below, no spots on sides. Reproduction: Spawns in late spring and early summer; male or both sexes guard large nest that can approach 3 ft. in diameter. Food: Aquatic invertebrates, fishes, and plants. Notes: White Catfish resemble Channel Catfish but have a wider head and mouth and lack a deeply forked tail and spots. White Catfish is a popular sport fish where abundant and most commonly found in the Coastal Plain.

Flat Bullhead
Ameiurus platycephalus
(right)

Snail Bullhead
Ameiurus brunneus
(not shown)

Flat Bullhead
Snail Bullhead
continued:

Size: 5–9 in.
Habitat: Warm rocky streams and rivers.
Abundance: Uncommon.
Status: See: Notes.

■ Flat Bullhead
■ Snail Bullhead

Description: Depressed frontal view, compressed toward tail; moderate to elongate profile; dark spot at base of dorsal fin; broad flat head; inferior mouth; moderate-sized eye; emarginate caudal fin; adipose fin separated from caudal fin (see p. 100); chin barbels pale (Flat Bullhead) or pigmented (Snail Bullhead); olive, brown, or gray above; pale yellow to green sides mottled in black to olive; white below. Reproduction: Spawn in early summer. Food: Aquatic invertebrates, fishes, and plants. Notes: Flat Bullhead is native to the Roanoke, Chowan, and Pee Dee drainages and possibly the James River. Snail Bullhead is found in the Dan River system, where it was introduced.

Yellow Bullhead
Ameiurus natalis

Size: 5–12 in.
Habitat: Sluggish water of warm rivers and creeks, ponds, lakes, and reservoirs.
Abundance: Uncommon to common.
Status: Native.

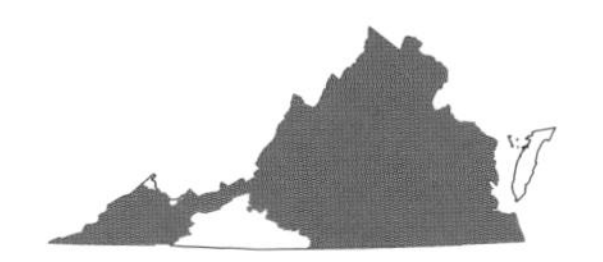

Description: Depressed frontal view, compressed toward tail; moderate to elongate profile; broad flat head; terminal or slightly subterminal mouth; small eye; emarginate caudal fin; adipose fin separate from caudal fin (see p. 100); dorsal fin lacks dark spot; pale chin barbels; olive to brown above and on sides, yellow or white below. Reproduction: Spawns during late spring and summer; nest builder. Food: Aquatic invertebrates, fishes, and plants. Notes: Yellow Bullhead is the most widely distributed bullhead in Virginia. It has an exceptional sense of smell and can recognize other Yellow Bullhead from their scent.

Brown Bullhead

Ameiurus nebulosus
(right)

Black Bullhead

Ameiurus melas
(not shown)

Size: 5–13 in.
Habitat: Ponds, lakes, reservoirs, and slow-moving streams and rivers.
Abundance: Rare to common.
Status: See Notes.

■ Brown Bullhead
■ Black Bullhead

Description: Depressed frontal view, compressed toward tail; moderate to elongate profile; broad flat head; slightly subterminal mouth; small eye; emarginate caudal fin; adipose fin separate from caudal fin (see p. 100); dorsal fin lacking dark spot; dark chin barbels; a pale vertical bar on caudal-fin base present (Black Bullhead) to absent (Brown Bullhead); olive, brown, black, or gray above and on sides, yellow or white below; mottling occasionally present. **Reproduction:** Spawn during summer; nest builder. **Food:** Aquatic invertebrates, fishes, and plants. **Notes:** Brown Bullhead is native to Atlantic Slope drainages in Virginia, with nonnative records in the New and Tennessee drainages. Black Bullhead is rare in Virginia, despite a large native range in the United States and southern Canada.

Flathead Catfish

Pylodictis olivaris

Size: 16 in. to 3 ft.
Habitat: Large warm streams, large rivers, lakes, and reservoirs.
Abundance: Rare to common.
Status: Native.

Description: Round to depressed frontal view, compressed toward tail; moderate profile; broad flat head; terminal mouth; small to moderate eye; emarginate tail; adipose fin separate from caudal fin (see p. 100); black, brown, or olive above and on sides, yellow below; fins yellow with brown mottling, upper lobe of caudal fin white when small.

Flathead Catfish
continued:

Other names: Appaloosa, Mudcat, Shovelhead Cat.

Reproduction: Spawns in summer; builds nests in cavities cleared of debris; nest and fry are guarded by male. Food: Aquatic insects and crustaceans as young, mostly fishes as adult. Notes: In Virginia, the Flathead Catfish is native to the Big Sandy, New, and Tennessee drainages but has been widely introduced elsewhere in the state. The Virginia record Flathead Catfish (68 lb., 12 oz.) was caught in May 2018 from Lake Smith.

Margined Madtom

Noturus insignis

Size: 3–7 in.
Habitat: Riffles, runs, and pools of medium streams to large rivers.
Abundance: Common.
Status: Native.

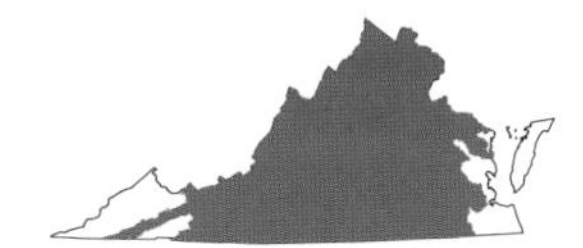

Description: Round frontal view, compressed toward tail; moderate to elongate profile; large eye; truncate tail; adipose fin connected to caudal fin (see p. 100); dark outline on the anal, caudal, and dorsal fins; brown, olive, green, or pale yellow above and on sides with yellow cast, white below; sometimes dark spots on sides (upper Dan River system). Reproduction: Spawns in early summer; nests under rocks in slower moving water near riffles. Food: Aquatic insects and crustaceans. Notes: Margined Madtom is native to the Atlantic Slope drainages and the New River. It is popular Smallmouth Bass bait, which may explain its introduction to the North Fork Holston River.

Orangefin Madtom

Noturus gilberti

Size: 2–3 in.
Habitat: Riffles and runs of small to large streams.
Abundance: Uncommon to rare.
Status: Native.

Description: Round frontal view, compressed toward tail; elongate profile; broad flat head; subterminal or inferior mouth; small eye; truncate tail; adipose fin connected to caudal fin (see p. 100); dark blotch on base of dorsal fin; pale margin on caudal fin, wider on upper lobe; gray, olive, brown, or yellow above and on sides; yellow or gray below; often has pink or orange tone. Reproduction: Spawns in late spring or early summer, likely under rocks. Food: Aquatic insects. Notes: Orangefin Madtom is state threatened due to habitat degradation from increased fine sediment. The species has a small native range (upper Roanoke drainage) but was introduced in the upper James River system.

Stonecat

Noturus flavus

Size: 2–6 in.
Habitat: Riffles and runs of medium to large streams.
Abundance: Rare.
Status: Native.

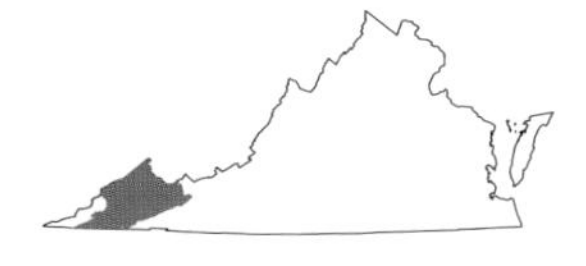

Description: Round frontal view, compressed toward tail; elongate profile; broad flat head; subterminal mouth; small or moderate-sized eye; truncate tail; adipose fin connected to caudal fin (see p. 100); pale crescent shape on nape and pale spot behind dorsal fin; dark blotch on base of dorsal and adipose fins; olive, brown, or yellow above and on sides, white below; pale fins. Reproduction: Spawn s in late spring or early summer; nests under rocks are guarded by the male. Food: Aquatic insects and crustaceans. Notes: Stonecat is rare in Virginia and only known from the Tennessee River drainage there.

Tadpole Madtom

Noturus gyrinus

Size: 2–5 in.
Habitat: Sluggish areas of warm, low-gradient streams and rivers.
Abundance: Rare to uncommon.
Status: Native.

Description: Round frontal view, compressed toward tail; moderate to deep profile; round head; rotund belly; terminal mouth; small eye; broad rounded caudal fin; adipose fin connected to tail (see p. 100); no dark margins on fins; gold or brown above, pale gold or brown on sides, with dark mid-lateral stripe, gold below. Reproduction: Spawns in late spring and summer; nests in sheltered areas, sometimes in cans and bottles. Food: Aquatic insects and crustaceans. Notes: Tadpole Madtom has one of the widest distributions of madtom species, occurring from Quebec and Manitoba, Canada, to southern Florida and eastern Texas.

Yellowfin Madtom

Noturus flavipinnis

Size: 3–4 in.
Habitat: Pools and backwaters of clear, flowing streams and larger rivers.
Abundance: Rare.
Status: Native.

Description: Round to depressed frontal view, compressed toward tail; moderate to elongate profile; broad, flat head; inferior mouth; large eye; rounded caudal fin; adipose fin connected to caudal fin (see p. 100); brown to gray or black saddles; saddle on the adipose fin; curved dark band on caudal-fin base; yellow brown to yellow gray above and on sides with pink hue, pale with pink hue below. Reproduction: Spawns from late spring to summer; male guards eggs under flat rocks in pools.

(continued)

Yellowfin Madtom
continued:

Food: Aquatic insects and crustaceans. **Notes:** Once thought to be extinct, Yellowfin Madtom occurs in the upper Tennessee River drainage in only a few rivers in Virginia. These geographically isolated populations are listed as a federally threatened species. The species has been successfully propagated and introduced to promote its recovery.

Mountain Madtom

Noturus eleutherus

Size: 2–3 in.
Habitat: Riffles and runs of warm streams and rivers.
Abundance: Common.
Status: Native.

Description: Round to depressed frontal view, compressed toward tail; moderate to elongate profile; broad flat head; inferior mouth; moderately large mouth; truncate tail; adipose fin connected to caudal fin (see p. 100); dark-brown saddles; dark bar on caudal-fin base; fairly thin caudal peduncle; dark brown above, lighter brown on sides with dark brown mottling, pale below with pink hue. **Reproduction:** Spawns during early summer; nests under rocks. **Food:** Aquatic insects and crustaceans. **Notes:** Mountain Madtom has a patchy native distribution within the Mississippi River basin and is Virginia's smallest madtom.

TROUTS
Family Salmonidae

Three species of trout (Salmonidae) inhabit the cold waters of montane Virginia. The Brook Trout is the Commonwealth's only native salmonid and holds the honor of being the State Freshwater Fish. Brook Trout is a circumpolar remnant of the Ice Age, usually found in high-gradient, forested locations in elevations typically higher than 1,000 feet above sea level. It is actually a char, one species of a branch of coldwater fishes that includes Lake Trout, Bull Trout, and Dolly Varden. Chars lack vomer teeth (see below) and are usually dark fishes with light-colored spots. Brown Trout was transported from Eurasia to North America in the 1800s and was stocked throughout western Virginia over time. Some streams have naturalized populations of Brown Trout. Brown Trout is favored for its sport, size, and flavor. Adult males can possess a hooked lower jaw, called a kype. Rainbow Trout comes from the mountainous Pacific Coast of North America. Its ability to produce high numbers of young fish in confined conditions make it a favorite of hatchery managers. All trout possess an adipose fin and develop parr marks as juveniles. A good time to observe Brook Trout is in October, during spawning season. Look for nesting pairs at the shallow end of pools before they turn into runs and riffles.

Vomerine teeth of trouts and chars:

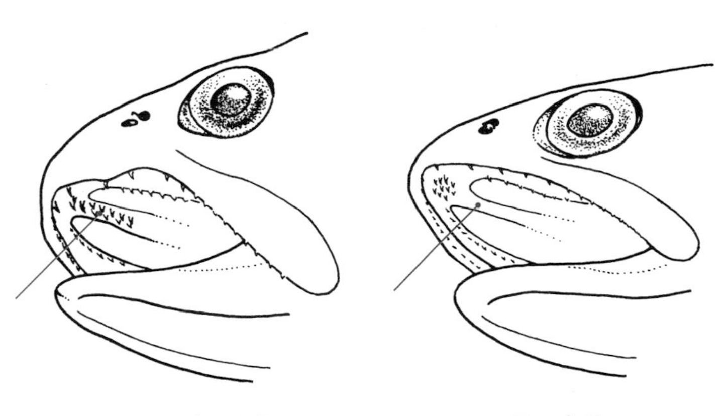

Brown Trout and Rainbow Trout: teeth present on vomer

Brook Trout: teeth absent on vomer

Brook Trout

Salvelinus fontinalis

Size: 4–7 in.
Habitat: Pools, runs, and riffles in montane streams.
Abundance: Abundant.
Status: Native.
Other names: Brook Char, Speckled Trout.

Description: Slightly compressed frontal view; moderate profile; terminal mouth; vomer teeth absent (see p. 109); adipose fin present; young have parr marks on sides; dorsal coloration dark with light vermiculation, red spots with blue halos on sides, orange underside; paired fins and anal fin reddish, trimmed in black and white. Reproduction: Spawns in fall when the female constructs and guards a gravel redd at pool tail. Food: Aquatic and terrestrial invertebrates. Notes: Male sometimes hybridizes with female Brown Trout, yielding "Tiger Trout." Brook Trout requires summer water temperatures below 71°F.

Brown Trout

Salmo trutta

Size: 8–22 in.
Habitat: Runs and deep pools in cold tailwaters, spring creeks, and montane streams.
Abundance: Not abundant.
Status: Introduced.

Description: Slightly compressed frontal view; moderate profile; terminal mouth; adipose fin present; vomer teeth present (see p. 109); young have parr marks on sides; dorsal color is dusky with black spots, red and black spots on sides, yellow underside; paired fins and anal fin yellow, pelvic fins sometimes trimmed in black and white, adipose fin sometimes red spotted, anal fin not spotted. Reproduction: Spawns in late fall on gravel bottoms with current. Food: Aquatic and terrestrial invertebrates when young; adults feed primarily on fishes such as sculpins, darters, and minnows. Notes: Has more environmental tolerance than Brook Trout and in proper conditions can grow to 25 in. or more.

Rainbow Trout

Oncorhynchus mykiss

Size: 8–12 in.
Habitat: Riffles and runs in cold tailwaters, spring creeks, and montane streams.
Abundance: Abundant.
Status: Introduced.

Description: Slightly compressed frontal view; moderate profile; vomer teeth present (see p. 109); adipose fin with black spots; young have parr marks on sides; dorsal coloration olive, light to heavy black spotting on sides, red or pink operculum and broad stripe along lateral line; dorsal fin, paired fins, and anal fins sometimes with white tips, anal fin, caudal fin, and dorsal fins heavily spotted. **Reproduction:** Spawns in late winter or early spring on gravel bottoms with current. **Food:** Aquatic and terrestrial invertebrates. **Notes:** More easily cultured than Brook or Brown trout and heavily stocked across western Virginia. Many naturalized populations compete with Brook Trout, particularly in southwest Virginia.

A Brook Trout in spawning color rises to take surface prey.

PIKES
Family Esocidae

The family Esocidae is distributed across the cooler portions of the Northern Hemisphere, including North America, Europe, and Asia. Fishes within the family are elongate with a duckbill-shaped snout, small scales, abdominal pelvic fins, no spines, and posterior-positioned dorsal and anal fins. Esocids use their keen vision to ambush prey from the cover of vegetation or submerged wood. Esocids primarily eat fishes but can opportunistically take snakes, frogs, ducks, and mammals such as muskrats. The larger pikes (Muskellunge and Northern Pike) are sought-after game fish due to their large attainable size, explosive strikes, and palatability. However, esocids are regarded as nuisance fishes by some anglers, who suspect they limit populations of other fishes. Rather, they are important predators and should be protected.

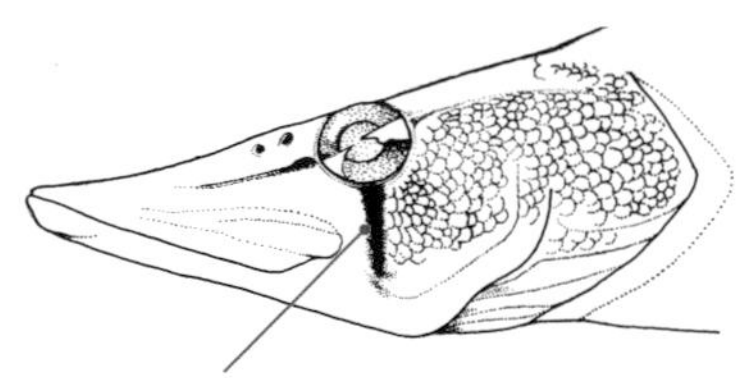
Chain Pickerel tear drop

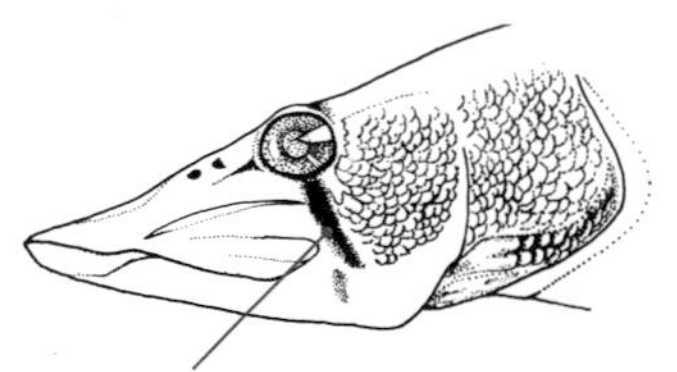
Redfin Pickerel tear drop

Northern Pike
Esox lucius

Size: 18 in. to 3 ft.
Habitat: Cool to warm small to large rivers and reservoirs.
Abundance: Rare.
Status: Introduced.

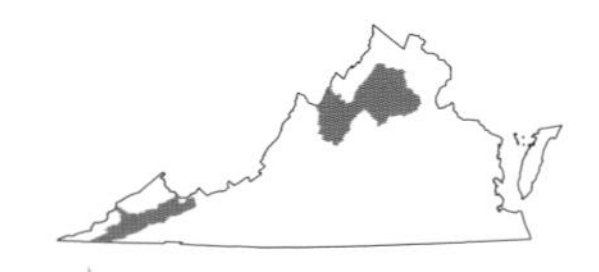

Description: Rounded frontal view, elongate profile; no tear drop below eye; blue to greenish gray or dark olive above and on sides, white below, irregular rows of white to yellow rectangular to oval-shaped spots over most of body; bronze, green, or red fins with black streaks or spots. **Reproduction:** Spawns in late winter or early spring; eggs are laid over vegetation in shallow water. **Food:** Fishes. **Notes:** Northern Pike is not native to Virginia but is stocked to provide a unique fishing opportunity. The state record is 31 lb., 4 oz. from Motts Run Reservoir, Spotsylvania County.

Muskellunge

Esox masquinongy

Size: 28 in. to 4 ft.
Habitat: Warm medium to large rivers and reservoirs.
Abundance: Uncommon.
Status: Introduced.
Other name: Musky.

Description: Round frontal view, elongate profile; no tear drop below eye; green to brown above, light green to silver on side, white below; dark bars or spots sometimes on sides; fins often red in color, with dark spots or streaks. Reproduction: Spawns in spring; spawning sites are ends of pools near riffles in streams or in shallow vegetated areas of lakes. Food: Fishes. Notes: Muskellunge are native to the Great Lakes and Mississippi River drainages but are believed to be introduced in Virginia based on historical records. Strong Musky populations in the James and New rivers provide excellent fishing opportunities.

Chain Pickerel

Esox niger

Size: 14–22 in.
Habitat: Warm creeks, rivers, ponds, and swamps.
Abundance: Uncommon.
Status: Native.
Other names: Chainsides, Jackfish.

Description: Round frontal view, elongate profile; long snout; dark vertical tear drop (see p. 112); dark green to brown above, yellow to light green on sides, with dark-green chain-like markings, white below; ventral fins colorless. Reproduction: Spawns in late winter and early spring; eggs laid on vegetation and debris. Food: Fishes and aquatic invertebrates. Notes: Chain Pickerel is the most common and widespread member of the family Esocidae in Virginia. In early spring, small juveniles can be seen lying motionless in inches of water, camouflaged as twigs or pine needles.

Redfin Pickerel

Esox americanus

Size: 5–8 in.
Habitat: Warm ponds, lakes, and streams.
Abundance: Common.
Status: Native.

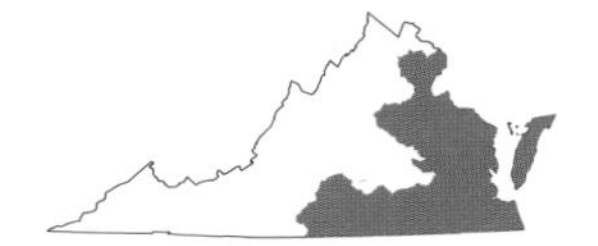

Description: Round frontal view, elongate profile; comparatively short snout; tear drop with slight backward slant (see p. 112); olive to brown above, dark vertical or slanted bars or blotches separate white to yellow bars on sides, white below; fins lack dark markings and may have red tint. Reproduction: Spawns in late winter or early spring; eggs are broadcast in shallows. Food: Fishes and aquatic invertebrates. Notes: The eastern subspecies *E. a. americanus* is found in Virginia and along much of the eastern seaboard, whereas the Grass Pickerel *E. a. vermiculatus* is found in the Mississippi River and Great Lakes drainages.

Chain Pickerel

MUDMINNOWS
Family Umbridae

The family Umbridae includes seven species of freshwater fishes, with only one inhabiting waters of the Commonwealth. Mudminnows are distributed across the Northern Hemisphere, including Eastern Europe, and from Alaska and Manitoba to Florida in the United States and Canada. Mudminnows are relatively small fishes with moderate to elongate bodies, rounded or truncate caudal fins, posteriorly positioned dorsal fins, abdominal pelvic fins, and small mouths. Many scientists believe that fishes from Umbridae should be classified within the family Esocidae, based on anatomical and genetic similarities. Members of the family are adapted to survive in either cold or warm waters that may be unsuitable for other fishes. Mudminnows can also survive in waters with low oxygen levels by using their gas bladder as a lung.

Eastern Mudminnow
Umbra pygmaea

Size: 2½–3½ in.
Habitat: Warm small slow-moving streams and rivers, ponds, swamps, and ditches.
Abundance: Uncommon.
Status: Native.

Description: Round frontal view, moderate profile; thick bodied; rounded caudal fin; black sunglass-shaped bar at base of caudal fin; brown to green above; light horizontal lines on sides; white to yellow below. **Reproduction:** Spawns in early spring; nest sites are on the bottom of algae mats or under loose rocks. **Food:** Aquatic invertebrates and plants. **Notes:** Eastern Mudminnow is distributed primarily on the Coastal Plain from New York to Florida. It is known to hide in cover during the day and feed at night.

PIRATE PERCH
Family Aphredoderidae

The Pirate Perch family consists of a single living species, the Pirate Perch. It is easily distinguishable by its large head, deep body, and a vent opening (anal-genital pore) in the throat region. Pirate Perch inhabits forested lowland areas of the Atlantic Slope, Gulf Slope, Mississippi Valley, and Great Lakes. The family is one of eight families of freshwater fishes endemic to North America. Pirate Perch is capable of chemical camouflage such that prey are unable to detect its presence via chemoreception.

Pirate Perch
Aphredoderus sayanus

Size: 2–5 in.
Habitat: Vegetated lowland streams, rivers, ponds, and backwaters.
Abundance: Uncommon.
Status: Native.

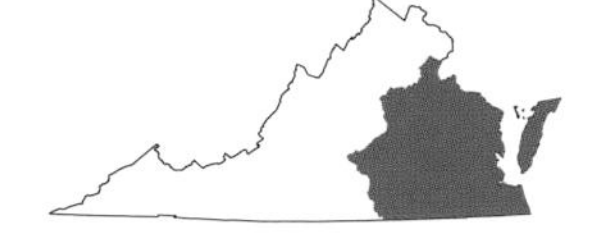

Description: Compressed frontal view; moderately deep profile; single dorsal fin; large mouth with projecting lower jaw; serrated preopercle; coloration is mostly dark brown or olive gray with black speckles and a narrow vertical dark bar on caudal-fin base and under the eye; young are dark, almost black; breeding adult may be violet or purple, nonbreeding adult is pinkish with dark-olive pigments. **Reproduction:** Spawn in underwater tree-root masses. Deposits eggs deeper by using tunnels created by dobsonfly larvae and salamanders. **Food:** Nocturnal feeder on a variety of insects, worms, glass shrimp, and small crayfish. **Notes:** Ichthyologist Charles Abbot coined its common name, Pirate Perch. The genus name *Aphredoderus* translates to "excrement throat." Species name *sayanus* is a tribute to naturalist Thomas Say, and "anus" translates to "belonging to." Although *sayanus* is the Latinized version of Say, legions of students remember the name by reciting "Say Anus" or "Say Anus Under Throat."

CAVEFISHES
Family Amblyopsidae

Cavefishes, or amblyopsids, are small fishes adapted to life in darkly stained surface waters, springs, and caves. Most species are cave dwellers with small ranges. All species in this family occur only in regions of eastern and central North America that were never glaciated. Amblyopsids are small (1–3 inches), with elongate bodies and specialized traits associated with subterranean or dark habitats. The body has extensive sensory pores, the eyes are small or nonfunctional, and the nasal openings are enclosed in tubes or bordered by flaps. The anal opening is close to the head, a rare trait shared only with the Pirate Perch. The only family member that occurs in Virginia is the Swampfish, *Chologaster cornuta*, a surface dweller. There are currently between seven and nine species of Amblyopsidae, all of which are geographically isolated from the Swampfish and at least partially cave adapted. A newly discovered cavefish, the Hoosier Cavefish, exists only in southern Indiana.

Swampfish
Chologaster cornuta

Size: 1–1½ in.
Habitat: Acidic blackwater swamps, sloughs, and streams with dense aquatic vegetation and coarse woody debris.
Abundance: Rare or uncommon.
Status: Native.

Description: Fusiform with flattened head, small eyes, upturned mouth, no pelvic fin, and a rounded caudal fin; brown coloration dorsally and creamy yellow belly with three dark longitudinal stripes on each side. Reproduction: Spawning occurs in March and April when Swampfish is one year in age. No one has observed spawning behavior so the purpose of the vent location remains a mystery. Food: Nocturnal or crepuscular feeders on small invertebrates. Notes: Young are born with the vent (i.e., anal-genital pore) positioned just anterior to the anal fin, and it migrates forward as the Swampfish matures. Mature male possesses a strange heart-shaped appendage on the snout; its function is still unknown. Swampfish is most easily collected with a dip net and adaps well in dimly lit aquariums with peat moss added to increase acidity.

NEW WORLD SILVERSIDES

Family Atherinopsidae

New World Silversides inhabit both marine and fresh waters of North, Central, and South America. All species are small, very elongate fishes with body length five to seven times the maximum body depth. There are 110 species in 13 genera in the Atherinopsidae, including the familiar California Grunnion, *Leuresthes tenuis*, Atlantic Silverside, *Menidia menidia*, Rough Silverside, *Membras martinica*, and the Inland Silverside, *Menidia beryllina*.

Brook Silverside

Labidesthes sicculus

Size: 3–4 in.
Habitat: Near surface of calm, open-water areas.
Abundance: Rare.
Status: Native.

Description: Compressed frontal view, elongate profile; flattened head, with large, beak-like terminal mouth; lateral silver band is broader anteriorly and is underlain by black pigment; body, opercle, and underside of the head are silvery white with iridescent blue green; appears transparent with pale shades of olive on the back and upper sides. Dorsal scales are clearly outlined. **Reproduction:** Spawns in May–June in shallow waters along the shoreline. Breeding silversides swim in a darting fashion toward the bottom, extruding eggs and sperm on their descent. Eggs have a long gelatinous filament that is used to attach to lake bottoms or rooted vegetation. After hatching, the young silversides make a wiggling swimming motion as they attempt to reach the water surface. **Food:** Planktonic crustaceans, small flying insects, and immature insects. **Notes:** Brook Silverside is well camouflaged in near-surface waters. A quick acceleration and a snap of the jaw result in rapid prey capture. It is often seen jumping out of water seeking flying insects. This Silverside is important forage for piscivorous fishes, mudpuppies, water snakes, and turtles.

TOPMINNOWS
Family Fundulidae

As the name implies, Topminnows are small, surface-dwelling fishes. The greatest diversity of this family occurs in the lowlands of eastern and central North America. Topminnows are also commonly referred to as killifish. Most Topminnows have a flattened head and protrusible mouth that is terminal or superior. They have teeth on the jaws as well as on pharyngeal bones. Most species show sexual dimorphism and brilliant male breeding colorations. Topminnows are adapted for life at the surface in calm, shallow margins. Because of their schooling habits, Topminnows are easy to catch with dip nets, seines, or cast nets. There are five species of Topminnows in Virginia's fresh waters.

Mummichog
Fundulus heteroclitus

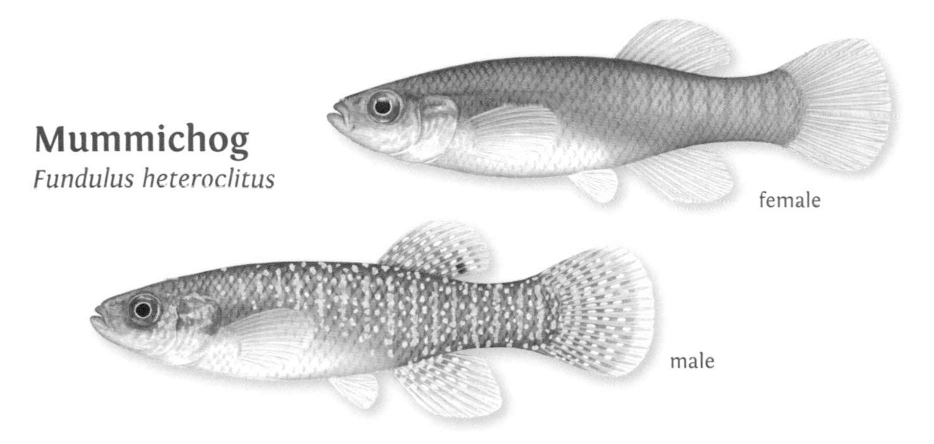

Size: 2–4 in.
Habitat: Marshes, canals, and tidal brackish creeks and rivers.
Abundance: Abundant.
Status: Native.
Other name: Mud Minnow.

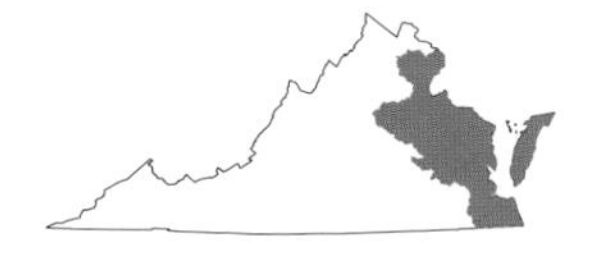

Description: Round frontal view, elongate profile; rounded head; rounded caudal fin; olive above, gray brown on side, white below; fins clear; adult male has yellow belly and bluish or gold barring and/or flecks on body and fins. Reproduction: Spawns in spring and summer; lays single eggs on vegetation and debris. Food: Insectivorous surface feeder. Notes: Mummichog is a Native American word meaning "going in crowds," referencing near-surface schooling behavior. This fish is extremely tolerant of temperature and salinity fluctuations and is often used in laboratory studies of physiology.

Banded Killifish

Fundulus diaphanus

Size: 2–4 in.
Habitat: Marshes, canals, and tidal brackish creeks and rivers.
Abundance: Abundant.
Status: Native.

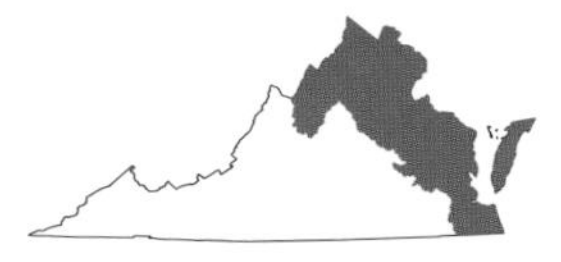

Description: Oval frontal view, elongate profile; more slender than Mummichog; flattened head; superior mouth; olive or silver overall, with clear fins; male has bluish or gold vertical bars along sides, whereas female has less-distinct dark vertical bars. Reproduction: Spawns in spring and summer, attaching single eggs to vegetation or debris. Food: Insectivorous surface feeder. Notes: The Banded Killifish is common in upland freshwater habitats. It is a popular bait fish and is known to survive for days at a time in wet leaves.

Speckled Killifish

Fundulus rathbuni

Size: 2–3 in.
Habitat: Slow runs and pool margins.
Abundance: Uncommon.
Status: Native.

Description: Oval frontal view, elongate profile; olive or tan overall with small black speckles on body (female) or head (male); dorsal and anal fins of male long and pointed, shorter and rounded on female. Reproduction: Spawns in spring and early summer. Food: Insectivorous surface feeder. Notes: This species is at the edge of its range in Virginia, occurring primarily in the piedmont of North Carolina.

Northern Studfish

Fundulus catenatus

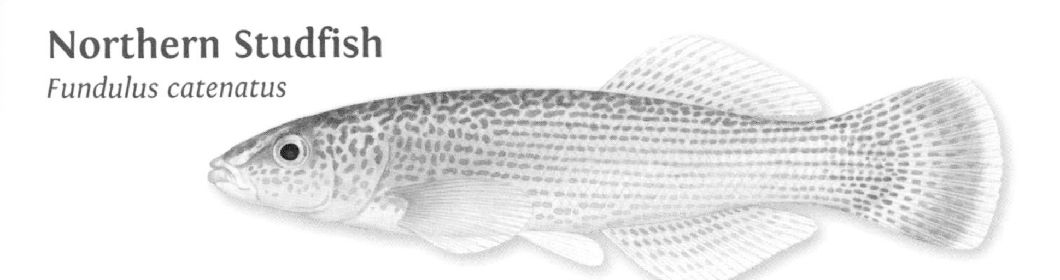

Size: 2–5 in.
Habitat: Shallow, sandy backwaters, pools, and creek margins.
Abundance: Uncommon.
Status: Native.

Description: Oval frontal view, elongate profile; upturned mouth; base color silvery with horizontal lines or rows of spots and from above, a bright silvery spot at base of dorsal fin; breeding male spectacular, with bluish body, red spots, and sometimes black-and-orange bands on caudal fin; male (even nonbreeding) has elongated, pointed dorsal and anal fins. **Reproduction:** Spawns in late spring and summer over clean sand and gravel substrates in shallow water. **Food:** Insectivorous surface feeder. **Notes:** This species can be spotted along sandy margins by looking for a bright silver spot moving just below the water's surface. When stationary, this species has a curious tendency to curl its tail toward one side.

Breeding male Northern Studfish with brilliant turquoise, red, and orange coloration.

Lined Topminnow

Fundulus lineolatus

Size: 2–3 in.
Habitat: Near shore or submerged and emergent vegetation in stained swamps.
Abundance: Uncommon or locally common.
Status: Native.

Description: Oval frontal view, elongate profile; flattened back, large fins, and upturned mouth; coloration olive yellow, often with coppery iridescence; female with horizontal black lines down body and often with pink-orange cheeks; mature male with gold-green hues, distinct vertical bars on sides, and slightly larger fins; both sexes have distinct silvery spot on top of head, making them highly visible from above water. Reproduction: Spawns in spring and summer, attaching eggs to vegetation. Food: Surface feeder on insects and planktonic organisms. Notes: This attractive species is one of the most distinct and visible fishes inhabiting the dark waters of the Coastal Plain and is enjoyable to observe from the comfort of shore.

LIVEBEARERS
Family Poeciliidae

The family Poeciliidae is found in fresh and brackish waters in North and South America as well as some regions in Africa and Madagascar. The family consists of approximately 274 species, including many popular aquarium fishes such as the guppies, mollies, swordtails, and platyfishes. The family name is derived from the Greek *poikilos*, meaning "with different colors." Members of this family have internal fertilization and bear live young. Virginia has only one species, the Eastern Mosquitofish, *Gambusia holbrooki*. Distinguishing external characteristics include a small upturned mouth, flat head, single dorsal fin, rounded tail, and pelvic fins near the abdomen.

Eastern Mosquitofish
Gambusia holbrooki

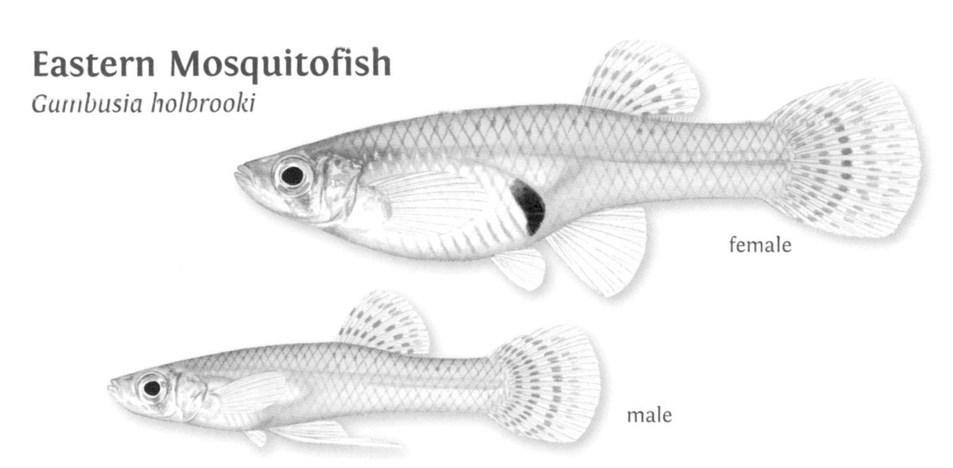

Size: ¾–2 in.
Habitat: Warm ponds, lakes, rivers, streams, marshes, and ditches.
Abundance: Common.
Status: Native.

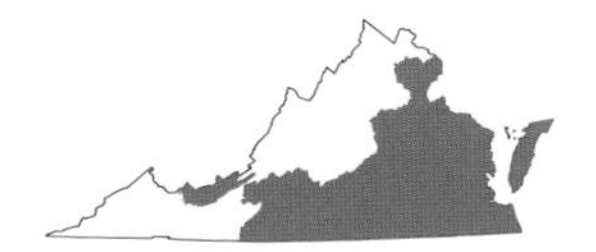

Description: Slightly compressed frontal view, moderate profile; small with transparent fins; upturned mouth; dusky to pale olive above, slight shimmer of green and purple on side; adult male with modified and elongate anal fin; adult female is larger, often with a dark gravid spot on abdomen. Reproduction: Spawns in spring and summer; livebearer. Food: Aquatic and terrestrial invertebrates and algae. Notes: Because of their name, mosquitofishes have been stocked worldwide to control mosquitoes. Research has indicated they have minimal effectiveness at controlling mosquitoes but can harm native fishes and amphibian species.

STICKLEBACKS
Family Gasterosteidae

The family Gasterosteidae has 19 species that are found in fresh water and marine environments in North America, Europe, and Asia. Four species occur in North America, with only one inhabiting the fresh waters of Virginia. Distinguishable external characteristics include sharp isolated dorsal spines in front of a rayed dorsal fin, bony plates instead of scales, and a thin, narrow caudal peduncle. Prominent bony plates along the sides are fewer and even lacking in freshwater sticklebacks. The family name is derived from the Greek *gaster*, "stomach," and *osteon*, "bone." They are a highly researched family because of their interesting behaviors and genetics.

Threespine Stickleback
Gasterosteus aculeatus

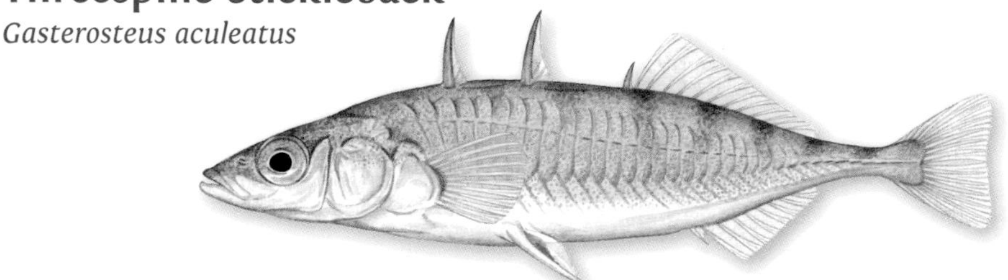

Size: 1½–3 in.
Habitat: Freshwater streams and brackish estuaries.
Abundance: Common.
Status: Native.

Description: Slightly compressed frontal view, moderate profile; three distinct spines on back; bony plates on back, sides, and belly; base of caudal fin is narrow; silver green, sometimes with brown mottling above, silver on sides and below; breeding male is brilliant blue green with red throat and blue eyes. **Reproduction:** Spawns in spring; male digs a small pit that is filled with plant material and guards and aerates the nest. **Food:** Aquatic invertebrates and plankton. **Notes:** Threespine Stickleback occurs on both the Atlantic and Pacific coasts of North America. Virginia is their southern-most range limit on the Atlantic coast. Species name *aculeatus* means "spined." Male courtship involves a dance with zigzag movements to entice a female to enter the nest.

SCULPINS
Family Cottidae

Sculpins, family Cottidae, inhabit both freshwater and marine habitats with an extensive range, including North America, Europe, and Asia. Cottidae is composed of at least 264 species, with 7 species known from Virginia waters. In addition, there may be several undescribed taxa within Virginia. Their tapering body, large head, large pectoral fins, and lack of a swim bladder are adaptations for benthic life in fast-flowing sections of streams and rivers. Other characteristics of the family include a large mouth, upward-oriented eyes, spiny projections on the gill cover, few to no scales, two dorsal fins, and a wide head. Sculpins are a popular fly pattern for fly fishermen, as sculpins often reside in trout streams. Species identification of sculpins is difficult, even for experienced fish biologists, and requires the use of a hand lens to examine chin pores, lateral line pores, preopercular spines, chin mottling, and the number of pelvic rays. Cottidae is from the Greek word *kottos*, meaning "head."

Sculpin median chin pore counts:

two median chin pores

one median chin pore

Sculpin chin pigmentation:

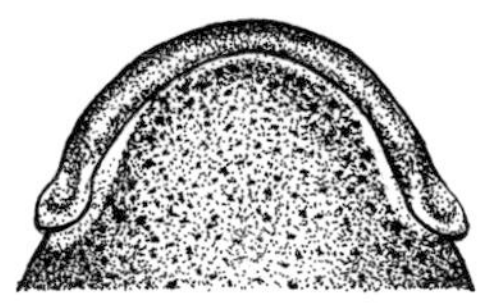

uniform pigmentation

mottled pigmentation

Sculpin caudal-fin base bands:

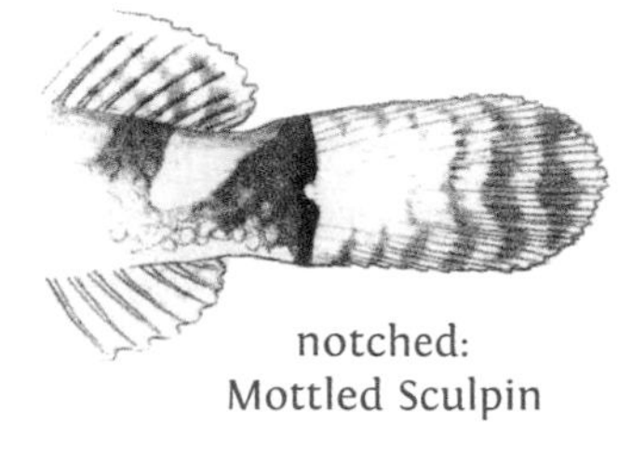

notched:
Mottled Sculpin

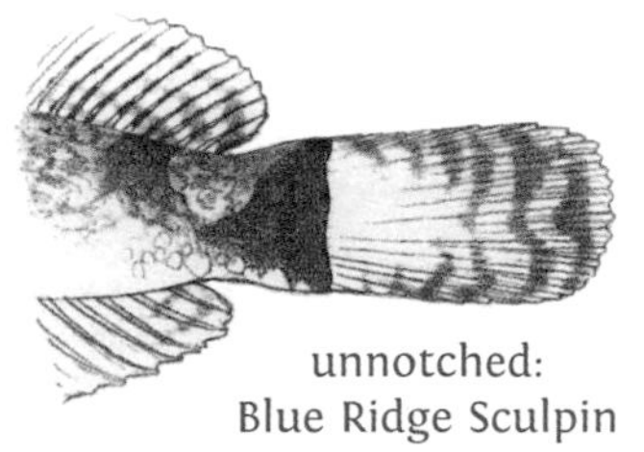

unnotched:
Blue Ridge Sculpin

Mottled Sculpin
Cottus bairdii (right)

Black Sculpin
Cottus baileyi
(not shown)

Blue Ridge Sculpin
Cottus caeruleomentum
(not shown)

Size: 2–3 in.
Habitat: Cool small to medium streams.
Abundance: Uncommon to common.
Status: Native.

- Mottled Sculpin
- Black Sculpin
- Blue Ridge Sculpin

Description: Rounded frontal view, compressed posteriorly; moderate profile; usually one (Blue Ridge Sculpin) or two (Mottled, Black, and Blue Ridge sculpins) pores in the middle of the chin; four pelvic rays; pectoral rays usually 14–15 (Blue Ridge and Mottled sculpins) or 15 (Black Sculpin); preopercular spines are reduced to short nubs (Mottled, Black, and Blue Ridge sculpins) to moderately strong (Blue Ridge Sculpin); chin uniformly pigmented; notches in dark band on both sides of caudal-fin base (Mottled Sculpin) or at least one side unnotched (Black and Blue Ridge sculpins) (see p. 125); incomplete lateral line; brown above, dark mottling on side and with two to four dark saddles on back, light below. **Reproduction:** Spawn in late winter (Black Sculpin) to spring (Mottled and Blue Ridge sculpins); males guard eggs in cavities under rocks or other debris. **Food:** Aquatic invertebrates. **Notes:** Sculpins from the Rappahannock River system are genetically distinct from both Blue Ridge Sculpin and Mottled Sculpin and require further study. Distinguishing between Black Sculpin and Mottled Sculpin is challenging, but fortunately their ranges do not overlap.

Slimy Sculpin

Cottus cognatus (right)

Potomac Sculpin

Cottus girardi

(not shown)

Size: 2–4 in.
Habitat: Cold small streams.
Abundance: Rare to abundant.
Status: Native.

Slimy Sculpin
Potomac Sculpin

Description: Rounded frontal view, compressed posteriorly; moderate profile; single pore in the middle of chin; three (Slimy Sculpin) or four (Potomac Sculpin) pelvic rays; pectoral rays usually 13–14 (Slimy Sculpin) or 15 (Potomac Sculpin); strong (Potomac Sculpin) or moderate (Slimy Sculpin) preopercular spines; mottled (Potomac Sculpin) or uniformly (Slimy Sculpin) pigmented chin (see p. 125); no (Potomac Sculpin) or two (Slimy Sculpin) black spots on first dorsal fin; incomplete lateral line; brown, olive, or tan above with four to five dark saddles (Potomac Sculpin), dark brown, gray, or green above, with dark mottling on side and two dark saddles (Slimy Sculpin). Reproduction: Spawn late winter (Potomac Sculpin) to early spring (Slimy Sculpin); cavity nesters. Food: Aquatic invertebrates. Notes: The generally smaller Slimy Sculpin is a northern species, with the southern extent of its range in Virginia's Potomac-Shenandoah system. Experts believe that Slimy Sculpin from the Potomac River system may be a distinct undescribed species, the "Checkered Sculpin." Potomac Sculpin can inhabit larger and warmer creeks and rivers.

Banded Sculpin

Cottus carolinae (right)

Kanawha Sculpin

Cottus kanawhae

(not shown)

Size: 3–4 in.
Habitat: Cold to warm small streams to large rivers.
Abundance: Rare to common.
Status: Native.

- Banded Sculpin
- Kanawha Sculpin

Description: Rounded frontal view, compressed posteriorly; moderate profile; two pores in the middle of chin; four pelvic rays; pectoral rays usually 16–17 (Kanawha Sculpin) or 15–17 (Banded Sculpin); strong preopercular spines; mottled chin (more intense in Banded Sculpin) (see p. 125); complete lateral line (sometimes incomplete in Kanawha Sculpin); brown, olive, or tan above, with four to five dark saddles; no black spots on first dorsal fin. **Reproduction:** Spawn in winter to early spring; females lay eggs on the underside of rocks that are guarded by the males. **Food:** Aquatic invertebrates and small fish. **Notes:** Kanawha Sculpin is endemic to the New River drainage, whereas Banded Sculpin inhabits a much larger geographic range from Alabama to Illinois and west to Oklahoma. Banded Sculpin is often broken into several subspecies across its range.

TEMPERATE BASSES
Family Moronidae

Three species and one hybrid represent this family in Virginia. Striped Bass and White Perch are anadromous, in that they spend most of their life in the Chesapeake Bay and Atlantic Ocean and migrate into rivers to spawn in fresh water. Some have been stocked elsewhere and can live their entire lives in fresh water. Striped Bass feed primarily on herring or shad and can grow to trophy size. White Bass is smaller as an adult than Striped Bass but when crossed with Striped Bass produces a desirable hybrid (Hybrid Striper). White Perch is found in brackish water and fresh water and is often preyed upon by larger fish. Temperate Basses are schooling fish. They are characterized by silver bodies, black stripes (except White Perch), and two dorsal fins—one soft and one spiny. All members of the family, especially Striped Bass, White Bass, and their hybrid, are desirable sport fish. Many of Virginia's reservoirs have been stocked with one or more of these species to develop pelagic fisheries.

Temperate Bass tongue tooth patches:

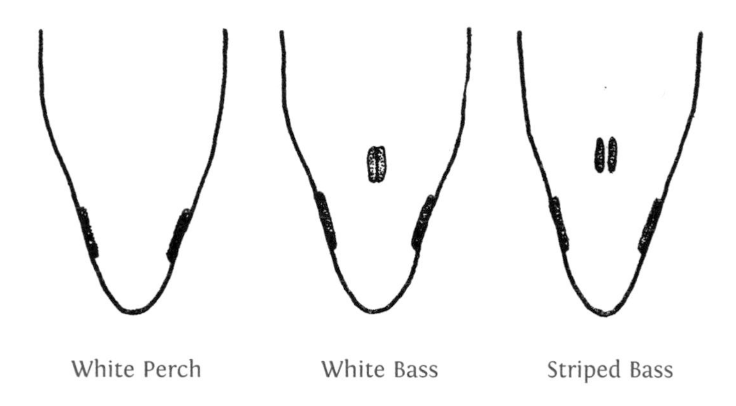

White Perch White Bass Striped Bass

White Perch

Morone americana

Size: 8–10 in.
Habitat: Estuaries, slow areas of small and large rivers, freshwater lakes.
Abundance: Abundant.
Status: Native.
Other names: Silver Perch, Silver Bass.

Description: Moderately compressed frontal view, deep profile; terminal mouth; median tooth patches on center of tongue absent (see p. 129); saw-like gill cover; back prominently arched; anal-fin spines stout, with first short; brassy to black above, silver-green sides with no stripes, white below. Reproduction: Spring pre-spawn migration; broadcasts eggs over sand or rocky bottom. Food: Crayfish, insects, and young fishes. Notes: Native to the coastal drainages but proliferates in reservoirs. Undesirable as a reservoir fish because diet overlaps with other game fish.

Striped Bass

Morone saxatilis

Size: 1–3 ft.
Habitat: Large rivers and reservoirs; can be anadromous or land locked.
Abundance: Common.
Status: Native.
Other names: Striper, Rockfish.

Description: Moderately compressed frontal view, moderate profile; terminal mouth; two distinct tooth patches on rear of tongue (see p. 129); anal-fin spines stout, graduated in length; dark olive to blue above, silver sides with unbroken black lateral stripes, white below. Reproduction: Spawning migration occurs in spring; broadcasts eggs over sand or rocky bottoms. Food: Alewife, herrings, shads. Notes: The largest inland Striped Bass was caught in Leesville Reservoir in 2000 and weighed 53 lb., 7 oz. Atlantic Ocean Striped Bass migrates seasonally from Maine to Florida.

White Bass
Morone chrysops

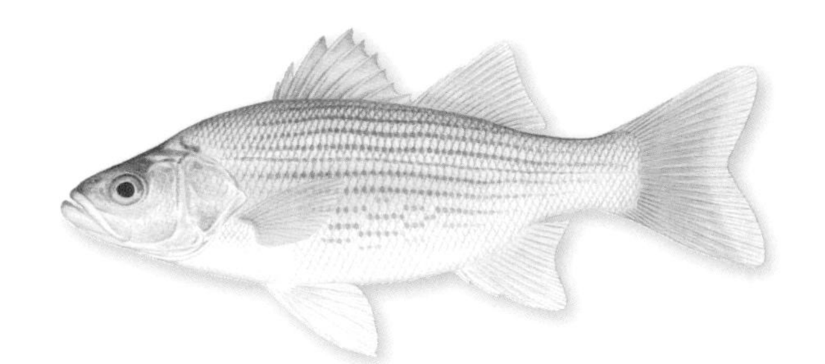

Size: 9–16 in.
Habitat: Large rivers and reservoirs.
Abundance: Uncommon.
Status: Native.
Other name: Sand Bass.

Description: Moderately compressed frontal view, deep profile; terminal mouth; single, oval tooth patch on rear of tongue (see p. 129); back prominently arched; anal-fin spines stout, graduated in length; blue to green or gray above, silver sides with faint, broken gray-brown stripes, white below. Reproduction: Spawning migration occurs in spring; broadcast spawner over sand or rock in spring. Food: Insectivorous when young, fishes and crayfish as adult. Notes: This schooling fish can provide hours of angling during a feeding spree. It is a strong fighter and excellent table fare. Native only to the Tennessee River drainage of Virginia.

Hybrid Striped Bass
Morone saxatilis female x
Morone chrysops male

Size: 12 in. to 2 ft.
Habitat: Reservoirs, lakes, and ponds.
Abundance: Common.
Status: Introduced.
Other names: Wiper, Palmetto Bass.

Description: Moderately compressed frontal view, deep profile; terminal mouth; two distinct tooth patches on rear of tongue; back prominently arched; anal-fin spines stout, graduated in length; dark olive to blue above, silver sides with strong, bold broken black stripes, white below. Reproduction: Functionally sterile. Food: Pelagic fishes. Notes: Hybrids are produced under hatchery conditions with sperm of male White Bass and eggs of female Striped Bass. Hybrid Striped Bass are desirable because they thrive in reservoirs, where cool summer habitat is limited for Striped Bass.

SUNFISHES
Family Centrarchidae

Centrarchid fishes are native to North American fresh waters and have been widely introduced. There are 38 species in 8 genera, all of which may be identified by a laterally compressed body, two connected spiny and soft dorsal fins, spines on anal fins, and pelvic fins in a thoracic position. The most diverse genus is *Lepomis*, commonly referred to as the sunfishes in recognition of their bright spawning colors. Virginia has 20 species including Mud Sunfish, Flier, 2 rock basses (*Ambloplites*), 3 banded sunfishes (*Enneacanthus*), 2 crappies (*Pomoxis*), 8 sunfishes (*Lepomis*), and 3 black basses (*Micropterus*). These fishes range widely in size from 3-inch Banded Sunfish to Largemouth Bass that may commonly exceed 30 inches.

Due to their popularity as a sport fish, sunfishes and black bass are widely introduced throughout North America and even other continents. The body form and habits all support sight feeding on a variety of crustaceans, insects, and small fishes. Males develop bright spawning coloration, establish territories, and build shallow, circular nests. Nests may be solitary or colonial, and males aggressively defend their nest, court females, and guard eggs and young. Closely related species have a tendency to hybridize.

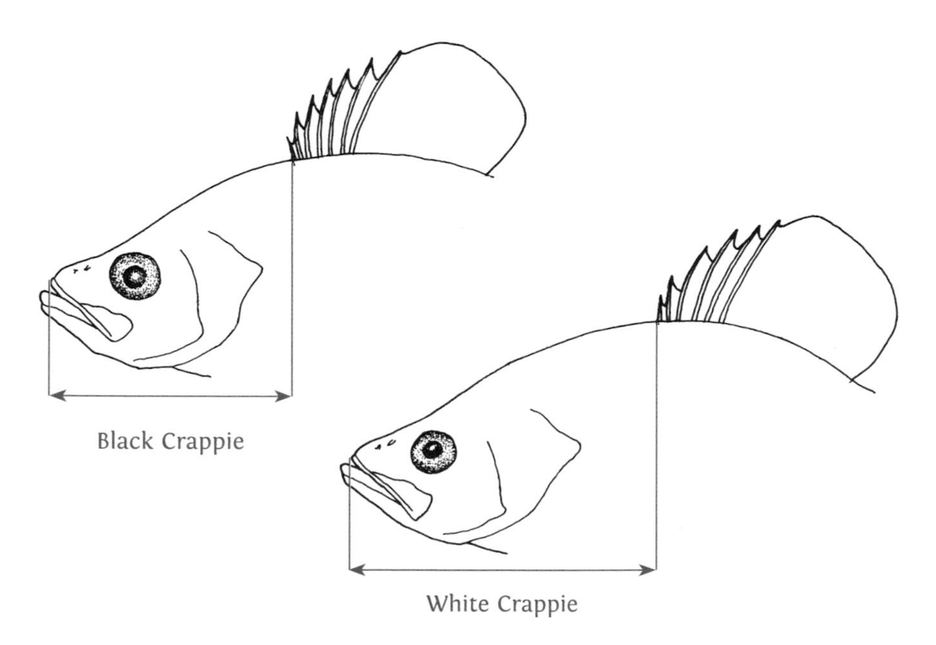

Opercular flap color patterns in the genus *Lepomis*:

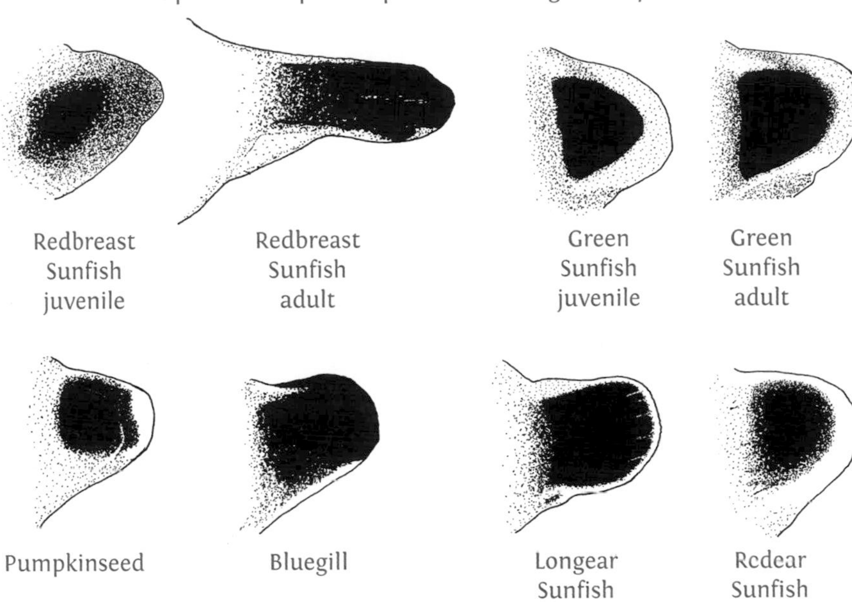

Mud Sunfish

Acantharchus pomotis

Size: 4–8 in.
Habitat: Sluggish, lowland acidic swamps, streams, and ponds with abundant cover and vegetation.
Abundance: Uncommon.
Status: Native.

Description: Laterally compressed frontal view with oblong profile; large, terminal mouth with large eye and short snout; three to four parallel brown to olive-black stripes across the face and three to four dark-brown stripes on side of body; opercular tab is short, with prominent spot with orange or light ventral and dorsal edges; rounded caudal fin.

(*continued*)

Mud Sunfish
continued:

Reproduction: Spawn from winter to early spring over small depression nests near shore. Food: Nocturnal feeder on crayfish, scuds, dragonfly larvae, beetles, and small fishes. Notes: Populations are possibly extirpated in New York and Pennsylvania. Vulnerable to disturbance of lowland habitats. Only member of the family Centrarchidae with cycloid scales. May be difficult to collect due to habitat requirements.

Rock Bass
Ambloplites rupestris

male

Size: 6–12 in.
Habitat: Clear streams in montane regions. Associated with shelter in pools and backwaters.
Abundance: Abundant in mountain streams.
Status: Native.
Other names: Branch Perch, Goggle Eye, Redeye, Rock Sunfish.

Description: Deep bodied, laterally compressed; large terminal mouth; head and dorsum dark olive or olive brown; prominent dark spots on the body scales form rows that are largest and darkest below lateral line; eye is partly or fully bright red; dark spot on edge of gill flap; fully scaled cheek; discernable dark spot on cheek; dark anal-fin margin; emarginate caudal fin; spawning male darker, and female more blotched. Reproduction: Male builds and defends a circular depression on coarse sand and gravel in April to June. Food: A generalist carnivore that feeds on crayfish, insects, and other fishes. Notes: Easy-to-catch and abundant sport fish in many of Virginia's favorite float-fishing rivers. Although the Rock Bass may seldom exceed 1 lb., it compensates with abundance and a lively fight on light tackle. Flesh is firm and delectable. Virginia state record is 2 lb., 2 oz. Rock Bass was widely stocked in Virginia.

Roanoke Bass
Ambloplites cavifrons

Size: 6–15 in.
Habitat: Rocky, gravel, and sandy pools in clear streams.
Abundance: Uncommon.
Status: Native.

Description: Laterally compressed frontal view, deep body profile; large terminal mouth; head and dorsum dark olive or brown; prominent dark spots on the body scales; pale spots on upper side and head; eye is partly or fully bright red; dark spot on edge of gill flap; unscaled or partly scaled cheek; discernable dark spot on cheek; anal fin with no dark margin; emarginate caudal fin. Reproduction: Male builds and defends a circular depression on coarse sand and gravel in April to June. Food: A generalist carnivore. Notes: Roanoke Bass is native only to Chowan and Roanoke of Virginia. There are historic records of 4 lb. Rock Bass that were most likely Roanoke Bass. Extirpated from historic range in Roanoke drainage above Roanoke, Virginia.

Flier

Centrarchus macropterus

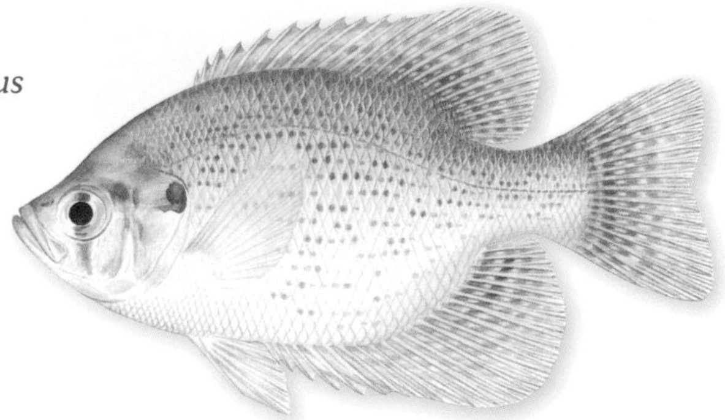

Size: 4–8 in.
Habitat: Acidic, well-vegetated swamps, ponds, and sluggish streams.
Abundance: Common.
Status: Native.

Description: Very compressed frontal view, deep body profile; base of dorsal fin long, with 11–12 (sometimes 13) spines and 12–15 soft rays; margin of dorsal fin broadly rounded posteriorly; slightly emarginate caudal fin; anal fin with 6–8 spines and 13–16 soft rays and its margin broadly rounded posteriorly; body olive green or brown and darker on back than belly; scales with brown spots forming longitudinal rows; wedge-shaped dusky bar through eye and cheek; smaller individuals have prominent orange-and-black ocellus (eye spot) on the posterior lower margin of dorsal fin. Reproduction: Male builds shallow nest in sand and gravel in March through April. Sometimes colonial breeder. Highest fecundity and smallest eggs of all centrarchid fishes. Food: Nocturnal feeder on crustaceans, small fishes, and aquatic insects such as caddisflies, mayflies, and midges. Notes: Attractive aquarium fish. Most similar to White Crappie and Black Crappie, which lack dark tear drop and rows of spots and have six to eight dorsal-fin spines. Small, but may be caught on artificial and live baits and often leaps out of the water; hence the name "flier."

Blackbanded Sunfish

Enneacanthus chaetodon

Size: 2–2½ in.
Habitat: Thickly vegetated acidic ponds, swamps, and pools and backwaters of streams.
Abundance: Rare.
Status: Native.

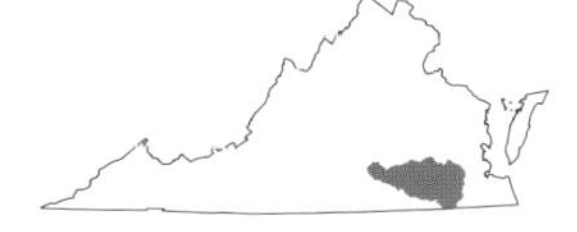

Description: Compressed frontal view, deep body profile; small terminal mouth; six bold black bars on sides—the first passes through the eye, the third extends dorsally through the anterior spiny dorsal fin and ventrally through the pelvic fin, and the sixth is fainter and extends through the caudal peduncle; back and upper side are dusky yellow gray, the side of body has yellow flecks; black spot on ear flap and black mottling on dorsal, anal, and caudal fins; reddish-orange streak along anterior spine and rays of pelvic fin. Reproduction: Male excavates and defends small depression nest in sand and gravel or creates hollows in algae or macrophytes. Male guards fertilized eggs and fry until they become free-swimming juveniles. Food: Opportunistic feeder on cladocerans, amphipods, midges (bloodworms), dragonfly nymphs, and caddisfly larvae. Notes: The "angelfish" of native fishes is of interest to aquarists. Cultured by fish breeders in Southeast Asia and Germany. Vulnerable or critically imperiled across most of its range due to drainage and development of wetlands and ponds. Endangered in Virginia. The name *Enneacanthus* reflects the "nine spines" in two of the three species. While adults are easily identified, the young can be very difficult to distinguish in the field.

Bluespotted Sunfish

Enneacanthus gloriosus

Size: 2–3½ in.
Habitat: Low-gradient, neutral to slightly acidic, stained and thickly vegetated swamps, ponds, and pools and backwaters of streams.
Abundance: Common.
Status: Native.

Description: Compressed frontal view, deep body profile; rows of bright blue (silver, gold, or green) spots along the side; black tear drop; rounded caudal fin; opercle and cheeks gold green with black opercular spot. Reproduction: Male excavates shallow depression nests in sand or beneath plants between May and September. Food: Opportunistic diurnal feeder on microcrustaceans and insects from water column and vegetation. Notes: Most common of the *Enneacanthus* in Virginia. Often confused with Banded Sunfish (see below). Adaptable and desirable aquarium fish if water is softened and acidified and provided with live foods or frozen bloodworms. The species name, *gloriosus*, means "glorious"—referring to beautiful coloration and markings.

Banded Sunfish

Enneacanthus obesus

Banded Sunfish

continued:

Size: 2–3½ in.
Habitat: Stained swamps, vegetated lakes and ponds, and pools and backwaters of streams.
Abundance: Rare.
Status: Native.

Description: Compressed frontal view, deep elongated profile; rounded caudal fin; body olive to olive green; mouth is supraterminal; five to seven olive or black lateral bars; black tear drop and pale or iridescent flecks on side; dorsal, caudal, and anal fins dark with rows of pale spots; opercle spot black with iridescent gold-green margin. Very similar to Bluespotted Sunfish. Reproduction: Male excavates shallow depression nests in sand or beneath plants between April and July. Food: Nocturnal feeder on microcrustaceans and insects. Notes: Popular aquarium fish, fares well in well-vegetated, softwater, acidic aquariums if given live foods or frozen bloodworms.

Black Crappie

Pomoxis nigromaculatus

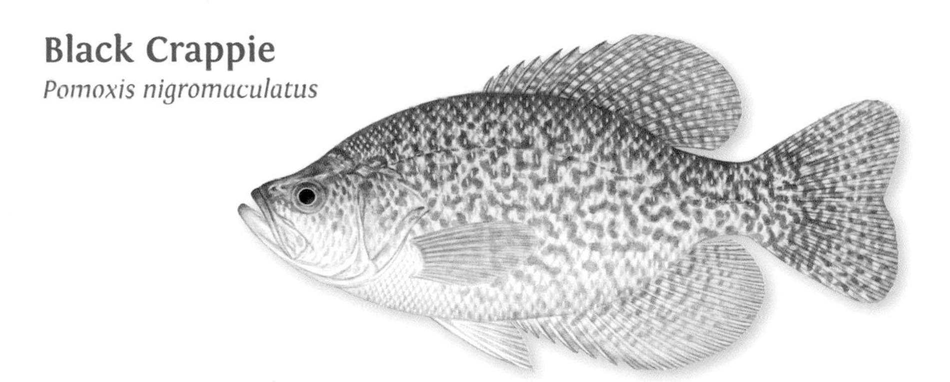

Size: 4–15 in.
Habitat: Ponds, lakes, rivers, and reservoirs associated with aquatic vegetation or woody cover.
Abundance: Common.
Status: Introduced.
Other names: Calico Bass, Oswego Bass, Papermouth, Slabs, Speckled Perch, Specks, Strawberry Bass.

Description: Very compressed frontal view, deep body profile; snout slightly upturned; mouth supraterminal, oblique, with protruding lower jaw; large eye; back is black or olive; long dorsal and anal fins with wavy patterns; dark olive to black scattered markings on side. Reproduction: Spawns in excavated nests in early spring in shallow water, often in colonies. Food: Feeds at dawn and dusk on minnows and aquatic invertebrates.

(*continued*)

Black Crappie
continued:

Notes: Similar to White Crappie but has seven to eight dorsal spines compared with five to six dorsal spines in White Crappie and a shorter distance from the snout to origin of dorsal fin (see p. 132). Very popular pan fish that will hit small jigs or live minnows. Pronounced "CROP-ee" throughout most of its range and "CRAP-ee" in parts of the South.

White Crappie
Pomoxis annularis

Size: 4–15 in.
Habitat: Ponds, lakes, and reservoirs associated with aquatic vegetation or woody cover.
Abundance: Common.
Status: Native.
Other names: Calico Bass, Goldring, Papermouth, Sac-au-lait, Silver Perch, Specks.

Description: Very compressed frontal view, deep body profile; snout slightly upturned; mouth supraterminal and oblique, with protruding lower jaw; large eye; back is olive; long dorsal and anal fins with wavy lines; six to ten vertical olive or black chain-like bars on a white or silver body. **Reproduction:** Spawns in excavated nests in early spring in shallow water, often in colonies. **Food:** Feeds at dawn and dusk on minnows and aquatic invertebrates. **Notes:** Chain-like bars and mottling pattern is often faint in specimens from turbid waters. Aggregations in spawning areas often popular locations for spring crappie anglers. Similar to Black Crappie but has five to six dorsal spines and a longer distance from the snout to origin of dorsal fin (see p. 132). Pronounced "CROP-ee" throughout most of its range and "CRAP-ee" in parts of the South.

Redbreast Sunfish

Lepomis auritus

Size: 4–9 in.
Habitat: Pools of small to medium-size rivers, usually associated with cover.
Abundance: Common.
Status: Native.

Description: Compressed frontal view, deep body profile; dark olive above and on sides with yellow flecks and rows of red-brown or orange spots on upper side and orange spots on lower side; white or orange belly; wavy, narrow blue lines radiate from mouth across sides of snout onto cheek and opercle; opercular flap of adult is long, narrow, and flexible, with black posterior margin that is not bordered above or below; shorter opercular flap in juvenile (see p. 133). Pectoral fin is short and rounded, tip not reaching past eye when bent forward. Reproduction: Nest builder in spring and summer. Food: Opportunistic feeder on insects and small fishes. Notes: Most similar to Longear Sunfish and Dollar Sunfish, but Redbreast Sunfish has longer ear flap with no posterior light margin and no blue marbling or spots on side. Popular sport and food fish on light tackle in streams and rivers.

Green Sunfish

Lepomis cyanellus

male

Size: 4–8 in.
Habitat: Streams, rivers, and shallow margins of lakes, ponds, and reservoirs.
Abundance: Locally common.
Status: Native.

Description: Slightly compressed frontal view, less deep body profile than other sunfishes; large terminal mouth is slightly oblique; upper jaw extends well beyond anterior edge of eye; blue green above and on side; green to blue wavy lines on snout, cheek, and opercle; opercular flap short, with pale margin and black center in juvenile and adult (see p. 133); body scales with dark scale margins and pale centers that may connect as broken stripes; pectoral fin is short and rounded, tip usually not reaching eye when laid forward across cheek; white or yellow belly; median fins are dusky with yellow, orange, or whitish margins, which are brighter in breeding male; dark spot usually present on posterior base of dorsal and anal fins. Reproduction: Nest builder in spring and summer, often in colonies. Food: Solitary ambush predator that feeds on a variety of larger invertebrates and fishes. Notes: Highly successful, aggressive fish that colonizes quickly after droughts and fish kills. Adult Green Sunfish are seldom confused with other species of *Lepomis*, all of which lack the yellow-orange fin margins. Hybrid between female Green Sunfish and male Bluegill is cultured and stocked for sport-fishing ponds because of its fast growth and aggressive behavior.

Pumpkinseed

Lepomis gibbosus

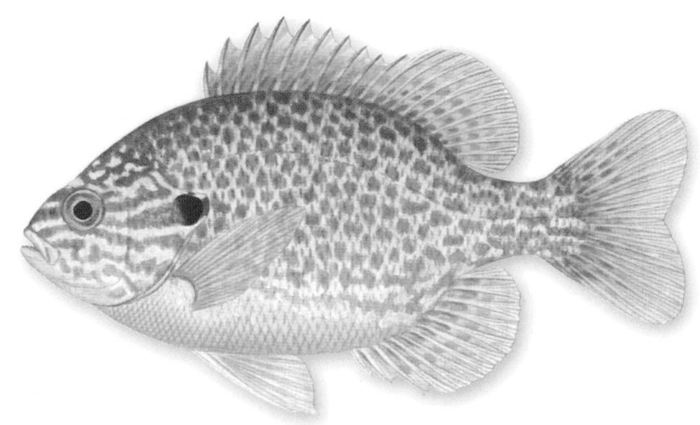

Size: 4–8 in.
Habitat: Vegetated streams, ponds, and reservoirs.
Abundance: Common.
Status: Native.

Description: Compressed frontal view, deep body profile; body with distinct pale yellow or orange spots encircled by dusky marks; wavy blue lines on the cheek and opercle; dorsal and anal fins mottled or spotted; opercular flap is short and stiff with a black center that is bordered by a semicircular spot on the posterior edge (see p. 133); spot may be white, pale yellow, or red; pectoral fin is long and sharply pointed; when bent forward, pectoral fin will reach beyond the front of the eye. Reproduction: Nest builder in spring and summer. Food: Pumpkinseed has specialized molariform teeth in the throat used for crushing snail shells, although it can feed opportunistically on a variety of aquatic and terrestrial invertebrates. Notes: Native only in Atlantic Slope drainages of Virginia but widely established via introductions. Pumpkinseed produces a chemical alarm substance that induces predator avoidance behavior in small Pumpkinseed. Most similar to Bluegill and Redear Sunfish, both of which lack wavy blue lines on cheek and opercle. The Redear Sunfish is further distinguished by the lack of mottling on dorsal and anal fins.

Warmouth

Lepomis gulosus

Size: 6–9 in.
Habitat: Vegetated lakes, ponds, swamps, reservoirs, and sluggish habitats in streams.
Abundance: Uncommon.
Status: Native.

Description: Slightly compressed frontal view, moderately deep body profile; large, terminal oblique mouth with lower jaw projecting slightly past the upper jaw; three to five dark-red bands radiate from the snout; dorsal profile of head is convex, indented over the eye; opercular flap is short, stiff, and black with paler, often red-tinged, border; coloration is olive brown with dark-brown mottling on back and side and dark spots and bands on fin; pectoral fin is short and round, usually not reaching the eye when laid forward; breeding male is boldly patterned with a red-orange spot at the base of the second dorsal fin and black pelvic fins. Reproduction: Nest builder near cover from spring to summer. Food: An omnivore that includes fishes, crayfishes, insects, and snails in its diet. Notes: Not well known to anglers in Virginia but does take a variety of baits and lures. Most common in Chowan and Dismal Swamp.

Bluegill

Lepomis macrochirus

Bluegill

continued:

Size: 4–12 in.
Habitat: All types of lacustrine habitats and pools of rivers.
Abundance: Abundant.
Status: Native.

Description: Compressed frontal view, deep body profile; small strongly oblique mouth; large black spot at posterior end of soft dorsal fin; pectoral fin is long and pointed; opercular flap is moderate or long and flexible with black margins (see p. 133); juveniles have pale silver or olive coloration and are usually marked with vertical chain-like bars; adults have two blue streaks from the chin to the edge of the gill cover (hence common name), a yellow-orange breast, and darker body coloration, with vertical bars often present. Reproduction: Nest builder in spring and summer, often in colonies. Food: Adaptable predator that may feed on zooplankton, small invertebrates often in association with aquatic vegetation, or terrestrial insects. Notes: Most common sunfish of Virginia and frequently stocked in sport-fishing ponds. Frequently overcrowded and stunted when underfished. State record is 4.5 lb.

Dollar Sunfish

Lepomis marginatus

(not shown)

Size: 1–4½ in.
Habitat: Swamp-like habitats and pools of lowland streams with abundant cover.
Abundance: Rare.
Status: Native.

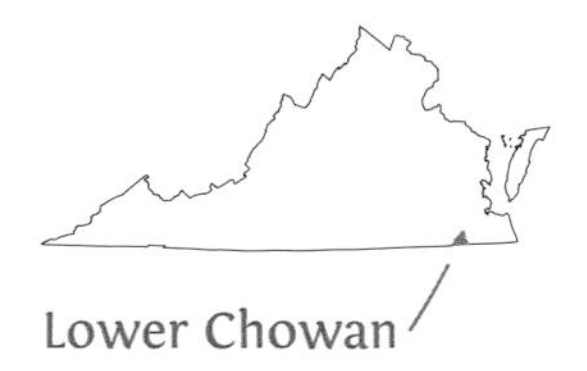

Description: Compressed frontal view, deep body profile; wavy blue lines on snout, cheek, and opercle; opercular flap is long and flexible, with a black center and white edges of equal width; pectoral fin short and rounded, not reaching the eye when laid forward; adults have an olive to brown body with scattered blue-green spots, red spots along the lateral line, and a yellow-orange breast and belly. Reproduction: Mature at 2½ in. and breeds in spring and summer in small nest depressions. Food: Opportunistic invertivore. Notes: Presence is an indicator of relatively undisturbed lowland and wetland ecosystems. Popular sunfish for large aquariums. Most similar to Longear Sunfish, but the distributions in Virginia do not overlap.

Longear Sunfish

Lepomis megalotis

Size: 4–7 in.
Habitat: Pools of headwaters, creeks, and small to medium-sized rivers.
Abundance: Uncommon.
Status: Native.

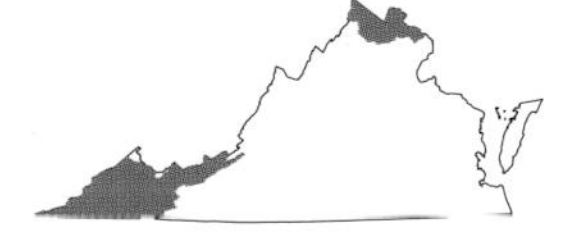

Description: Compressed frontal view, deep body profile; distinctive wavy blue lines on snout, cheek, and opercle; opercular flap is long and flexible, with a black center and white edges of equal width (see p. 133); pectoral fin is short and rounded, not reaching the eye when laid forward; adult is dark olive brown above, bright orange ventrally, and marbled and spotted with blue on the side. **Reproduction:** Nest builder in spring and summer, often in colonies. **Food:** Feeds on variety of aquatic and terrestrial invertebrates, often feeds on aquatic insects disturbed by feeding hog suckers. **Notes:** Longear Sunfish vigorously attacks a variety of baits and is frequently caught with spin and fly-fishing tackle. One of the most colorful of Virginia's sunfishes. This species is becoming increasingly rare in its native range due to competition with Redbreast Sunfish.

Redear Sunfish

Lepomis microlophus

Redear Sunfish
continued:

Size: 4–11 in.
Habitat: Clear, vegetated ponds, lakes, and reservoirs and sluggish pools and backwaters of rivers.
Abundance: Rare.
Status: Introduced.
Other names: Government Bream, Shellcracker.

Description: Compressed frontal view, deep body profile; short opercular flap with black center and bordered above and below with white margins and posteriorly with a prominent red or orange crescent (see p.133); coloration is light gold green above with many dark connected spots on the side; dorsal and anal fins typically clear, but some mottling may be present near fin bases; pectoral fin long and pointed. Reproduction: Nest builder in spring and summer, often near vegetation and in colonies. Food: Opportunistic omnivore specialized for feeding on snails and small mussels. Notes: Redear Sunfish resembles Bluegill and Pumpkinseed. It does not have wavy blue lines on the head or mottled dorsal and anal fins (Pumpkinseed) and does not have dark marking on second dorsal fin (Bluegill). It is widely stocked for sport fishing and eats snails and small bivalves, earning it the name "Shellcracker." Anglers catch Redear Sunfish with worms and other natural baits fished near the bottom.

Smallmouth Bass
Micropterus dolomieu

Size: 11 in. to 2 ft.
Habitat: Cool and warm streams and rivers with large gravel and cobble bottoms and riffles, pools, and runs.
Abundance: Common.
Status: Native.

Description: Slightly compressed frontal view, elongate body profile; large terminal mouth; upper jaw extends at least to below center of eye; connection between soft and spiny dorsal fins slightly notched and broadly connected; cheek with three dark-olive bars; dark brown with numerous bronze markings on scales; often with 8 to 16 indistinct vertical bars on yellow green to brown side; iris red.

(*continued*)

Smallmouth Bass *continued*:

Other names: Bareback Bass, Bronzeback, Bronze Bass, Brownie, Smallie.

Reproduction: Solitary nest builder near cover; spawns from April to June. Food: Opportunistic top carnivore that feeds from surface to bottom on fish, crayfish, and terrestrials. Notes: Most sought-after and valued sport fish in Virginia's rivers; state record of 8 lb., 1 oz. caught in New River. Oft-quoted description is "inch for inch and pound for pound the gamest fish that swims" (James A. Henshall, *Book of the Black Bass*, 1881). Appears similar to Spotted Bass, which has dark mid-lateral stripe and rows of dark spots along lower sides.

Smallmouth Bass sac fry.

Spotted Bass
Micropterus punctulatus

Size: 11–22 in.
Habitat: Low-velocity pools and cover in warm-water streams and impoundments.
Abundance: Uncommon.
Status: Native.
Other names: Kentucky Bass, Spot.

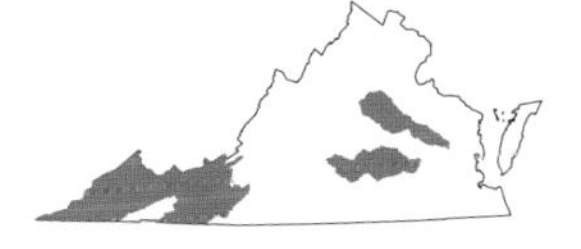

Description: Slightly compressed frontal view, elongate body profile; large terminal mouth; upper jaw extends little or not at all beyond posterior edge of eye; connection between soft and spiny dorsal fins slightly notched and broadly connected; rows of small spots on yellow-white lower sides arranged in horizontal lines; dark mid-lateral stripe formed from series of partly joined dark blotches along olive yellow side; caudal spot dark, especially in young specimens; tooth patch present on tongue. Reproduction: Solitary nest builder in spring. Food: Opportunistic carnivore. Notes: Co-occurs with Smallmouth Bass and Largemouth Bass but seldom reaches trophy sizes in Virginia waters.

Largemouth Bass

Micropterus salmoides

Size: 12 in. to 2.6 ft.
Habitat: Impoundments, lakes, ponds, swamps, backwaters, and pools of creeks and rivers.
Abundance: Common.
Status: Native.
Other names: Bigmouth Bass, Black Bass, Bucketmouth, Florida Bass, Florida Largemouth, Gilsdorf Bass, Green Bass, Green Trout, Oswego Bass, Potter's Fish, Widemouth Bass.

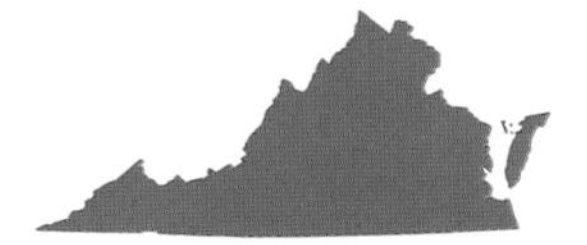

Description: Slightly compressed frontal view, elongate body profile; large terminal mouth; upper jaw extends beyond posterior edge of eye in adult; connection between soft and spiny dorsal fins deeply emarginated and almost separate; outline of spiny dorsal fin sharply angular, not curved; dark midlateral stripe formed from series of confluent dark blotches along olive-yellow side; tooth patch usually absent on tongue. Reproduction: Solitary nest builder in spring. Food: Opportunistic top carnivore that feeds primarily on fishes as adult. Notes: Most sought-after and valued sport fish in Virginia reservoirs and tidal rivers. This keystone species may have profound effects on prey fish and habitat use, and its harvest is regulated to enhance fishing. Can be distinguished from Spotted Bass by absence of teeth on tongue and size of mouth.

PERCHES
Family Percidae

Perches and darters are a northern-hemisphere family with at least 237 known species and 11 genera. While some species are of conservation concern because of their vulnerability to extinction, others are important for recreational and commercial fisheries. Most of the species are North American darters, and their phylogeny and species distributions are a subject of active research. All species in Percidae have ctenoid scales, two dorsal fins, pelvic fins in thoracic position, and spines on cheeks and gill covers. Size ranges widely, from the diminutive 1-inch Golden Darter *Etheostoma denoncourti* to the 41-inch, 25-pound Walleye, *Sander vitreus*. Five genera (*Sander*, *Perca*, *Ammocrypta*, *Percina*, and *Etheostoma*) occur in Virginia. Darter species possess distinctive and beautiful color patterns that are pronounced in males during the breeding season. Conservation of the darters and preservation of their habitats should be a priority, as 44% of Percidae are imperiled.

Sauger
Sander canadensis

Size: 10–18 in.
Habitat: Warm medium to large rivers and reservoirs.
Abundance: Rare.
Status: Native.

Description: Rounded frontal view, elongate profile; pointed snout; forked caudal fin; large head and eye; large sharp teeth; bronze to brown above and on side, white below; several dark-brown or black saddles across back, extending down side; fin mostly clear, with yellow-and-black stippling and round black spots on first dorsal fin. Reproduction: Spawns in early to mid-spring; eggs broadcast in deep rocky runs. Food: Fishes and invertebrates. Notes: Sauger is a smaller relative of the Walleye that occurs less commonly in Virginia. Artificial hybrid of the two species (called Saugeye; see next page) is occasionally produced and stocked to provide fishing opportunities in small impoundments due to its fast growth rate.

Walleye

Sander vitreus

Size: 16 in. to 2.6 ft.
Habitat: Warm large rivers and reservoirs.
Abundance: Uncommon.
Status: Native.
Other names: Jackpike, Pike-perch, Walleyed Pike.

Description: Rounded frontal view, elongate profile; pointed snout; forked caudal fin; large head and eye; large sharp teeth; solid brown or bronze above and on side, white below; occasionally with faint mottling on side; fin color usually similar to body color, caudal fin typically yellow, with white tip on the lower lobe; first dorsal fin sometimes has black stippling (not round spots as in the Sauger). Reproduction: Spawns in early to mid-spring; eggs are broadcast in rocky runs and riffles. Food: Fishes and crayfish. Notes: Walleye is a popular game fish, both for sport and its mild, sweet flavor. The species is native to the Big Sandy, New, and Tennessee drainages and stocked widely elsewhere.

"Saugeye," *Sander vitreus* female x *Sander canadensis* male

Yellow Perch

Perca flavescens

Size: 5–12 in.
Habitat: Cool to warm streams to large rivers and reservoirs.
Abundance: Common.
Status: Native.
Other names: Raccoon Perch, Ringed Perch.

Description: Compressed frontal view, moderate profile; snout rounded; forked caudal fin; olive above, gold to greenish on side, white below; about seven dark vertical bars tapering ventrally; dorsal and caudal fins dusky; anal, pelvic, and pectoral fins often bright orange. Reproduction: Spawns in late winter; attaches strings of eggs to aquatic vegetation and debris. Food: Aquatic invertebrates and fishes. Notes: Yellow Perch is native to Atlantic Slope rivers but has been introduced elsewhere. An anadromous population in the Chesapeake Bay migrates into tributaries to spawn.

Sharpnose Darter

Percina oxyrhynchus

Size: 3–4¾ in.
Habitat: Warm medium streams to large rivers.
Abundance: Uncommon.
Status: Native.

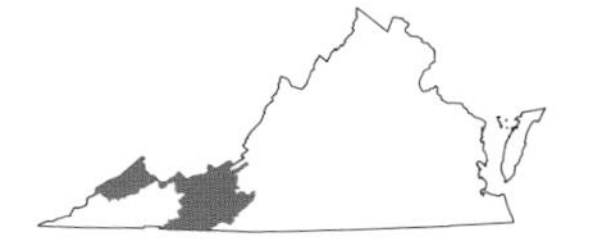

Description: Rounded frontal view, elongate profile; extremely long pointed head and snout; 9–12 light to dark round blotches on side connected by a thin, faded stripe; olive to yellow brown above, with wavy lines and small blotches on upper side, pale yellow fading to white below; bright orange band on dorsal fin of breeding male, more subtle in female; occasionally orange tint on caudal and second dorsal fin. Reproduction: Spawns in spring; buries eggs in gravel. Food: Aquatic invertebrates. Notes: Sharpnose Darter is infrequently seen and has a rather unusual range, including the upper New River and Big Sandy drainages in Virginia.

Logperch
Percina caprodes (right)

Blotchside Logperch
Percina burtoni
(below right)

Chesapeake Logperch
Percina bimaculata (extirpated)

Size: 4–6 in.
Habitat: Medium streams to large rivers.
Abundance: Uncommon.
Status: Native.

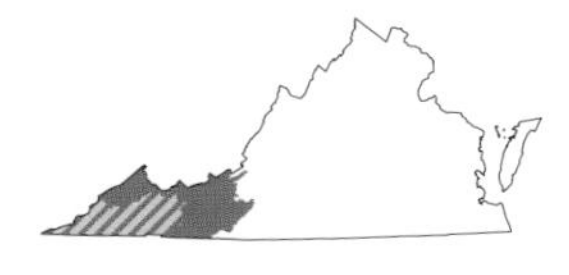

■ Logperch
■ Blotchside Logperch

Description: Rounded frontal view, elongate profile; long, bulbous snout; base color tan, pale olive green to light brown above, many dark vertical stripes (Logperch) or oval blotches (Blotchside Logperch) on tan to greenish side, white below; fins mostly clear, with black speckling or rays; males with orange-and-black bands on dorsal fins (Blotchside Logperch). **Reproduction:** Spawn in spring; bury eggs in fine gravel. **Food:** Aquatic invertebrates. **Notes:** The extirpated Chesapeake Logperch once occurred in the Potomac River but now occurs only in parts of Pennsylvania.

Roanoke Logperch
Percina rex

Size: 4–7 in.
Habitat: Warm medium streams to large rivers.
Abundance: Rare to uncommon.
Status: Native.

Description: Rounded frontal view, moderate to elongate profile; long, bulbous snout; base color tan, pale olive green to light brown above, many vertical bars on tan to greenish side, white below; fins mostly clear with black speckling or rays; male with orange-and-black bands on dorsal fins. **Reproduction:** Spawns in spring; buries eggs in fine gravel. **Food:** Aquatic invertebrates. **Notes:** All logperches display the unique behavior of flipping rocks by using their large snouts in search of food. Roanoke Logperch is listed as federally endangered.

Sickle Darter

Percina williamsi

Size: 2½–3½ in.
Habitat: Warm medium streams to large rivers.
Abundance: Rare.
Status: Native.

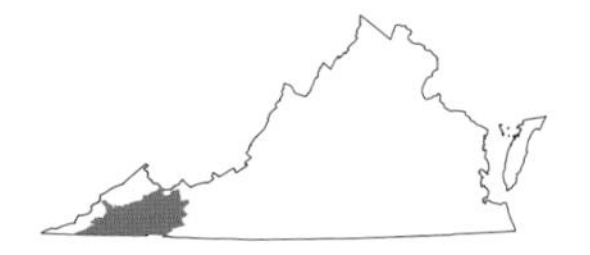

Description: Rounded frontal view, elongate profile; long head and pointed snout; 8–14 connected blotches on side; tan above, with wavy dark lines from tip of nose to base of caudal fin, cream or white below blotches; fins clear, with first dorsal fin having a black marginal band. Reproduction: Spawns in spring; buries eggs in fine gravel. Food: Aquatic invertebrates. Notes: Prior to 2007, Sickle Darter was considered *Percina macrocephala*. Sickle Darter is threatened in Virginia and found only in the Clinch and Holston drainages.

Appalachia Darter

Percina gymnocephala (right)

Blackside Darter

Percina maculata (not shown)

Size: 2–4¼ in.
Habitat: Warm small streams to large rivers.
Abundance: Rare to uncommon.
Status: Native.

Description: Rounded frontal view, moderate profile; moderately rounded snout; caudal fin slightly emarginate; six to nine large black blotches connected by a mid-lateral stripe; tan to olive above, upper side with wavy dark bands, white to yellow on lower side and below; dark tear drop below eye; fin typically clear or slightly yellow; breeding male with green to blue iridescence on cheek and opercle; two yellow spots (Appalachia Darter) or single dark blotch (Blackside Darter) on base of caudal fin.

(*continued*)

Appalachia Darter
Blackside Darter
continued:

- Appalachia Darter
- Blackside Darter

Reproduction: Spawn in spring; bury eggs in gravel. Food: Aquatic invertebrates. Notes: Blackside Darter ranges widely throughout the eastern and central United States, but only a few records are known from Virginia in the Big Sandy drainage, where they are possibly extirpated. Appalachia Darter is endemic to the New River drainages in North Carolina, Virginia, and West Virginia.

Stripeback Darter
Percina notogramma (right)

Dusky Darter
Percina sciera
(not shown)

Size: 2½–4 in.
Habitat: Warm medium streams and rivers.
Abundance: Rare to uncommon.
Status: Native.

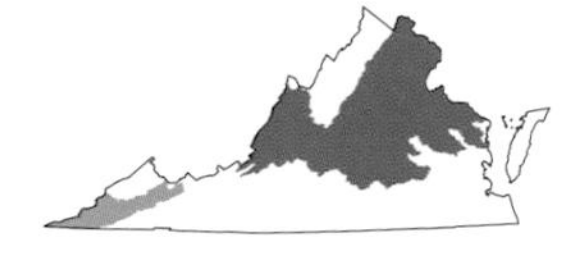

- Stripeback Darter
- Dusky Darter

Description: Rounded frontal view, moderate profile; moderately rounded snout; caudal fin slightly (Stripeback Darter) to fully (Dusky Darter) emarginate; 6–8 (Stripeback Darter) to 8–12 (Dusky Darter) interconnected, dark, oval blotches on side; olive to tan above, pale to dusky yellow on side, and tan to white below; dark tear drop present in Stripeback Darter and absent in Dusky Darter; Dusky Darter with three dark spots on caudal-fin base, with lower two fused to form irregular blotch, not present in Stripeback Darter; posterior blotches of Dusky Darter breeding male extend to back; dorsal fin dusky to olive brown (Dusky Darter) with dark crescents anteriorly (Stripeback Darter); juvenile Stripeback Darter with smaller rectangular blotches on side and dark blotch on base of caudal fin. Reproduction: Spawn in spring; bury eggs in fine gravel. Food: Aquatic invertebrates. Notes: Stripeback Darter may be confused with the Shield Darter where the two species co-occur. Stripeback Darter has large oval blotches, compared to the rectangular to square blotches of the Shield Darter.

Shield Darter

Percina peltata (right)

Chainback Darter

Percina nevisense (not shown)

Size: 2–3¾ in.
Habitat: Warm medium streams to large rivers.
Abundance: Common to uncommon.
Status: Native.

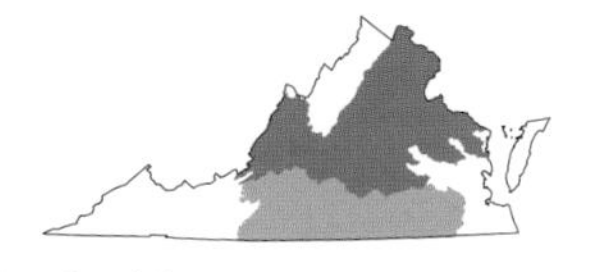

■ Shield Darter
■ Chainback Darter

Description: Rounded frontal view, slightly elongate profile (more slender in Chainback); moderately rounded snout; six to nine rectangular, dark lateral blotches, sometimes connected by narrow band; olive green to straw above; silver to gold to green on upper side, lighter on lower half of side and below; chainlike pattern down back when viewed from above (more distinct in Chainback); fins typically clear to yellow or olive, first dorsal fin has black crescents on membranes. **Reproduction:** Spawn in spring; bury eggs in sand and gravel near riffles. **Food:** Aquatic invertebrates. **Notes:** Chainback Darter is found in the Roanoke and Chowan drainages. It is now a separate species after having long been considered a subspecies of Shield Darter.

Roanoke Darter

Percina roanoka (right)

Piedmont Darter

Percina crassa (not shown)

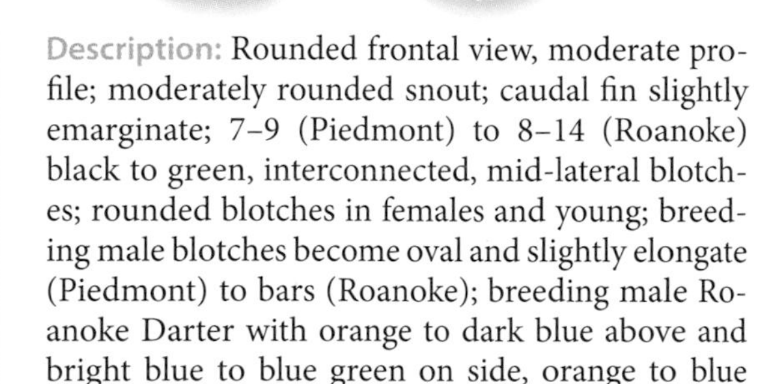

male

Size: 1¾–3½ in.
Habitat: Warm medium streams to large rivers.
Abundance: Rare to common.
Status: Native.

Description: Rounded frontal view, moderate profile; moderately rounded snout; caudal fin slightly emarginate; 7–9 (Piedmont) to 8–14 (Roanoke) black to green, interconnected, mid-lateral blotches; rounded blotches in females and young; breeding male blotches become oval and slightly elongate (Piedmont) to bars (Roanoke); breeding male Roanoke Darter with orange to dark blue above and bright blue to blue green on side, orange to blue below; breeding male Piedmont Darter with yellow

(*continued*)

Roanoke Darter
Piedmont Darter
continued:

■ Roanoke Darter
■ Piedmont Darter

above, on side, and below; dorsal fin with bright yellow (Piedmont) to orange band (Roanoke). **Reproduction:** Spawn in spring; bury eggs in sand and gravel mix with cobble and boulder. **Food:** Aquatic invertebrates. **Notes:** Roanoke Darter is native to the Roanoke River and was introduced in the New and James drainages. Virginia is the northern extent of the Piedmont Darter range in the Pee Dee drainage.

Gilt Darter

Percina evides

male

Size: 1¾–3¾ in.
Habitat: Warm medium streams to large rivers.
Abundance: Common.
Status: Native.

Description: Rounded frontal view, moderate profile; moderately rounded snout; caudal fin slightly emarginate; seven to nine dark-green to black mid-lateral blotches extending over back; base color of female and juvenile olive above, yellow on upper side, white on lower side and below; fins mostly clear; breeding male with iridescent blue-green head, side and below varying from blue green to orange red; dorsal fin dark, with burnt orange band; all other fins dark with orange streaks. **Reproduction:** Spawns in spring and early summer; buries eggs in sand-gravel among cobble and boulder. **Food:** Aquatic invertebrates. **Notes:** The Gilt Darter is typically less benthic than most other darters, often preferring to forage while hovering a few inches above the stream bottom.

Tangerine Darter

Percina aurantiaca

male

Size: 4–7 in.
Habitat: Medium streams to large rivers.
Abundance: Uncommon.
Status: Native.
Other name: River Slick.

Description: Rounded frontal view, elongate profile; moderately rounded snout; lacks the black tear-drop bar under the eye; breeding male black and olive above, with dark blotches on upper side (often with green iridescence), bright orange on lower side and below; color pattern continues onto head; juvenile and female light olive above and with dark conjoined blotches on upper side, with a line of smaller dark spots in between, creamy white (juvenile) to yellow (female) on lower side and below; fins clear with orange band on dorsal fins; lower fins of breeding male dusky to bluish; caudal and dorsal fins bright orange, sometimes with diffuse dusky black. Reproduction: Spawns in spring; buries eggs in gravel in fast currents. Food: Aquatic invertebrates. Notes: Large specimens are big enough to roll gravel in search of prey. Breeding male Tangerine Darter is breathtakingly beautiful; a band of fluorescent orange spans from chin and across lower cheek and belly. Tangerine Darter is endemic to the upper Tennessee drainage and is one of the largest darter species.

Channel Darter

Percina copelandi

Size: 1½–2½ in.
Habitat: Warm small to medium rivers.
Abundance: Rare.
Status: Native.

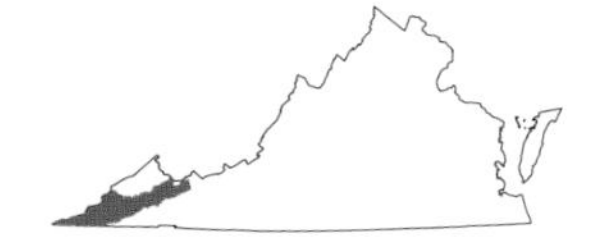

Description: Rounded frontal view, elongate profile; slender, moderately rounded snout; tan to light brown with dark speckling above, circles or rectangular blotches mid-laterally, white below; fins mostly clear, with some dusky patterning on dorsal fin. **Reproduction:** Spawns in spring and early summer; buries eggs in fine sand and gravel substrates in swift currents. **Food:** Aquatic invertebrates. **Notes:** Channel Darter is quite distinct from other *Percina* species and more closely resembles Johnny Darter and Tessellated Darter in both size and behavior.

Western Sand Darter

Ammocrypta clara

Size: 1½–2¾ in.
Habitat: Warm medium to large rivers.
Abundance: Rare.
Status: Native.

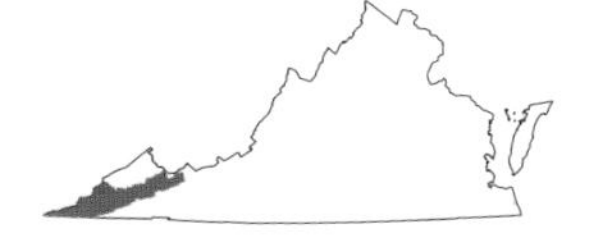

Description: Rounded frontal view, elongate profile; moderately long snout; transparent above and on side, opaque white below; thin gold stripe and 18–24 small dark blotches on back; thin gold stripe mid-laterally, with small spots; cheek with blue-green iridescence; fins clear. **Reproduction:** Spawns in summer; lays eggs in shallow riffles. **Food:** Aquatic invertebrates. **Notes:** As its name indicates, Western Sand Darter is highly associated with sandy habitats, where it is difficult to find in the rocky rivers where it is known to exist. Western Sand Darter is listed as threatened in Virginia.

Ashy Darter

Etheostoma cinereum

Size: 2–3 in.
Habitat: Warm small to medium rivers.
Abundance: Rare.
Status: Native.

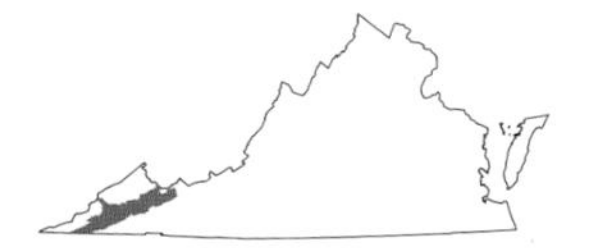

Description: Slightly compressed frontal view, moderate to elongate profile; long, pointed snout; second dorsal fin extremely large in male; tan or brown with dark, diagonal bands on side, sometimes with blotches; caudal and dorsal fins stippled with red; female with less color overall and smaller dorsal fins. Reproduction: Spawns in early spring; attaches eggs to substrate. Food: Aquatic invertebrates. Notes: Ashy Darter was presumed extirpated from Virginia until it was rediscovered in the mid-2000s. It is a species of special concern considered for federal listing under the Endangered Species Act.

Swannanoa Darter

Etheostoma swannanoa

Size: 2–3½ in.
Habitat: Cold to cool medium to large streams.
Abundance: Uncommon.
Status: Native.

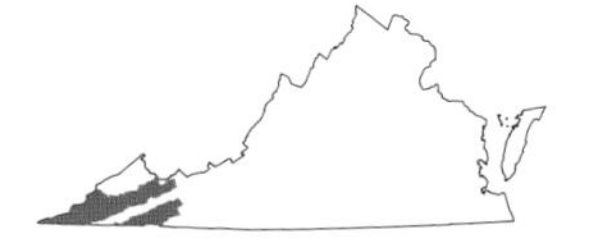

Description: Slightly compressed frontal view; moderate profile; short, blunt snout; extremely long pectoral fins; tan to brown to olive above, becoming lighter on side, white below; six dark-brown to black saddles; 8–11 dark blotches (female) to bars (male); horizontal rows of rusty brown to orange spots on body (male); fin color clear or slightly dusky (female) to turquoise coloration in pelvic, anal, and dorsal fins (male); breeding male has turquoise on fins and under head. Reproduction: Spawns in early to mid-spring; buries eggs in fine to pea-sized gravel. Food: Aquatic invertebrates. Notes: Swannanoa Darter has a disjunct distribution in the Tennessee drainage.

Candy Darter

Etheostoma osburni (right)

Kanawha Darter

Etheostoma kanawhae (not shown)

Variegate Darter

Etheostoma variatum (not shown)

male

Size: 2–4½ in.
Habitat: Cool to warm streams to medium rivers.
Abundance: Rare to uncommon.
Status: Native.

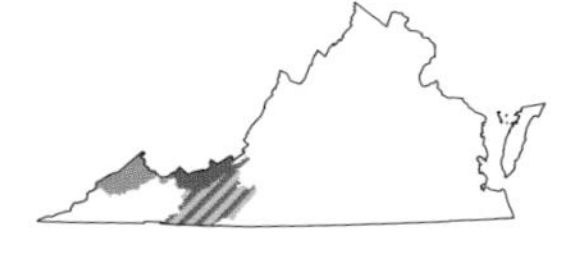

- Candy Darter
- Kanawha Darter
- Variegate Darter

Description: Rounded to slightly compressed frontal view, moderate to slightly elongate profile; moderately rounded snout; large brown saddles ranging from four (Variegate Darter), five (Candy Darter), to five to six (Kanawha Darter); yellow to olive above and on side, white below; breeding males with blue or blue-green and bright red alternating bars posteriorly that fade anteriorly (Variegate Darter and Kanawha Darter) or extend across entire side (Candy Darter); diffuse red or red horizontal stripe across belly connected to red bars (Candy Darter); females drab, with dark blotches and speckling; fin color mostly clear in females, dorsal fin of breeding males blue or blue green with red band. **Reproduction:** Spawn in late spring and early summer; bury eggs in gravel behind rocks in riffles. **Food:** Aquatic invertebrates. **Notes:** Breeding male Candy Darters are one of the most beautiful fishes in North America. Candy Darter is federally endangered, while Variegate Darter is protected in Virginia as endangered. Although not listed, Kanawha Darter is a New River endemic, known to inhabit streams draining the Blue Ridge physiographic province. Variegate Darter hybridizes with Candy Darter.

Greenside Darter

Etheostoma blennioides

Size: 3–5 in.
Habitat: Cool to warm streams to large rivers.
Abundance: Common.
Status: Native.

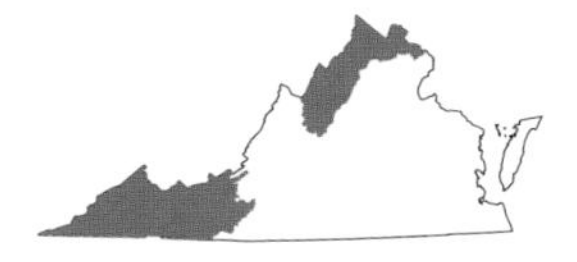

Description: Rounded frontal view, elongate profile; blunt snout; large pectoral fins; tan or olive above, olive on side, pale green below; six to seven saddles, dark speckling on upper side, and distinct U-shaped markings on lower half; dorsal fin often with orange band; breeding male becomes darker, with emerald-green bands on body and emerald-green or turquoise fins. Reproduction: Spawns in early to late spring; lays eggs on aquatic vegetation in moderate to fast currents. Food: Aquatic invertebrates. Notes: Greenside Darter has a wide distribution extending north to Canada and as far west as Oklahoma. It is thought to be introduced in the Potomac River system.

Banded Darter

Etheostoma zonale

Size: 1½–3 in.
Habitat: Warm medium to large rivers.
Abundance: Uncommon.
Status: Native.

Description: Rounded frontal view, elongate profile; slightly blunt snout; tan to opaque white above, on side, and below; brown mottling and dark bands faint (female) to dark green (male); green bands extend underneath belly; fins mostly clear (female) or faint green (male), orange and green bands on anterior dorsal fin; dark tear drop. Reproduction: Spawns from mid-spring to early summer; attaches eggs to vegetation in shallow water. Food: Aquatic invertebrates. Notes: Banded Darter is widespread and ranges north to New York and west to Minnesota and the Ozarks.

Snubnose Darter

Etheostoma simoterum

Size: 1½–2¾ in.
Habitat: Cool to warm small streams to large rivers.
Abundance: Common.
Status: Native.

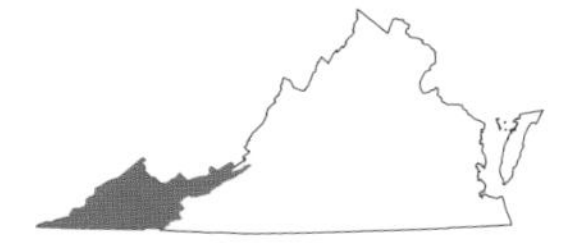

Description: Slightly compressed frontal view; moderate profile; very short, blunt snout; female and nonbreeding male tan to gray to green above and on side, pale yellow below; seven to nine saddles; side with seven to nine dark blotches, sometimes interconnected, speckling of brown (female) or solid orange stripe (male) on upper side, breeding male usually with orange on lower side and down belly; fin color clear with brown speckling (female) or orange speckling (red bands on male dorsal). Breeding male has bright, blue-tipped snout. Reproduction: Spawns in spring; female attaches eggs to stream bottom or vegetation. Food: Aquatic invertebrates. Notes: Snubnose Darter is native to the Tennessee River system in Virginia. It is possibly introduced in the Big Sandy and the New River drainage, where it is rapidly spreading.

Blueside Darter spawning pair.

Bluespar Darter

Etheostoma meadiae (right)

Blueside Darter

Etheostoma jessiae (below right)

Size: 1–2¾ in.
Habitat: Warm medium streams to large rivers.
Abundance: Rare to common.
Status: Native.

- Bluespar Darter
- Blueside Darter

Description: Slightly compressed frontal view; slightly elongate profile; snout moderate (Bluespar Darter) to slightly pointed (Blueside Darter); juveniles and nonbreeding adults olive to light brown above, pale yellow to tan on side, pale white below; speckled on back and a row of dark unconnected squares on side; fin color clear, with dorsal fin sometimes displaying faint orange margin; difficult to distinguish nonbreeding specimens; breeding males have orange flecks all over body and royal-blue blotches (Blueside Darter) or bars (Bluespar Darter) on side and cheeks; soft dorsal and anal fin blue near base; base of caudal fin blue or green, with bar extending from dorsal to ventral margin; dorsal fin has blue, white, orange, and blue bands from edge to base; bright orange spots present on soft dorsal and caudal fins of Blueside Darter but not Bluespar Darter. **Reproduction:** Spawn in spring; bury eggs in sand or fine gravel. **Food:** Aquatic invertebrates. **Notes:** Blueside Darter and Bluespar Darter are virtually colorless and perfectly camouflaged on a sandy river bottom most of the year, but males become extremely colorful during the short breeding season.

Riverweed Darter

Etheostoma podostemone (right)

Longfin Darter

Etheostoma longimanum
(not shown)

male

Size: 1–3 in.
Habitat: Warm small streams to medium rivers.
Abundance: Uncommon to common.
Status: Native.

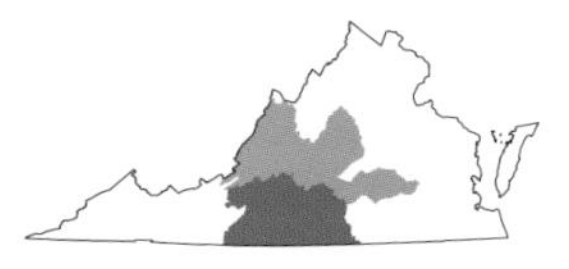

■ Riverweed Darter
■ Longfin Darter

Description: Rounded frontal view, elongate profile; moderately blunt snout; enlarged second dorsal fin; female and nonbreeding male tan to opaque white to brown above and on side with dark speckling on upper side, dark "XW" markings or separated blotches mid-laterally, tan to pale yellow below; fin color clear to yellow; breeding male gold, yellow, or orange with distinct dark saddles above and blotches mid-laterally; bright orange fins; orange spots on nearly all body scales, with many scales edged in iridescent blue purple (Riverweed Darter) or green to violet (Longfin Darter). Reproduction: Spawn in spring to early summer; attach eggs to the underside of rocks; eggs guarded by the male. Food: Aquatic invertebrates. Notes: These two species have very small ranges mostly within Virginia. Longfin Darter is endemic to the James and Riverweed Darter is endemic to the Roanoke River system.

Tessellated Darter

Etheostoma olmstedi (right)

Johnny Darter

Etheostoma nigrum
(not shown)

Size: 1–4 in.
Habitat: Warm small streams to large rivers.
Abundance: Common.
Status: Native.

Description: Rounded frontal view, elongate profile; moderately blunt snout; tan to opaque white above, on side, and below; black speckling on back and side, with many brown "XW" markings on side; fin color clear to slightly yellowish with dark, broken, wavy stripes; breeding males' head and fins dusky to black.

Tessellated Darter
Johnny Darter
continued:

■ Tesselated Darter
■ Johnny Darter

Reproduction: Spawn in spring. Males guard eggs until hatching. Food: Aquatic invertebrates. Notes: In Virginia, the Johnny Darter and Tessellated Darter have overlapping ranges, resulting in an intergrade zone for these closely related species. Distinguishing the two species is difficult, especially in areas where their ranges overlap.

Glassy Darter
Etheostoma vitreum

Size: 1–2½ in.
Habitat: Warm medium creeks to large rivers.
Abundance: Uncommon.
Status: Native.

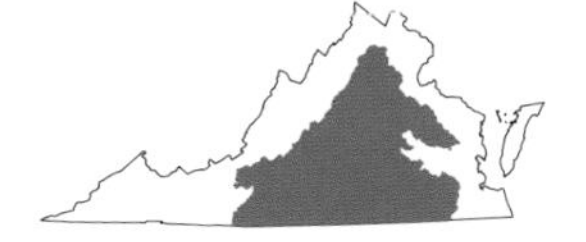

Description: Rounded frontal view, elongate profile; long, pointed snout; opaque white to tan to nearly translucent above and on side, white below; peppering of small black spots all over body; small black rectangles mid-laterally; fin color clear with slight patterning on fin rays; breeding male black to dusky. Reproduction: Spawns in early to mid-spring in groups; scatters eggs on stream bottom. Food: Aquatic invertebrates. Notes: Glassy Darter is one of the few darter species that lives on shifting sandy bottoms, where their translucent body provides excellent camouflage.

Bluebreast Darter

Etheostoma camurum
Syn.: *Nothonotus camurus*
(right)

Greenfin Darter

Etheostoma chlorobranchium
Syn.: *Nothonotus chlorobranchius*
(not shown)

male

Size: 2–4 in.
Habitat: Cool to warm medium creeks to large rivers.
Abundance: Rare to common.
Status: Native.

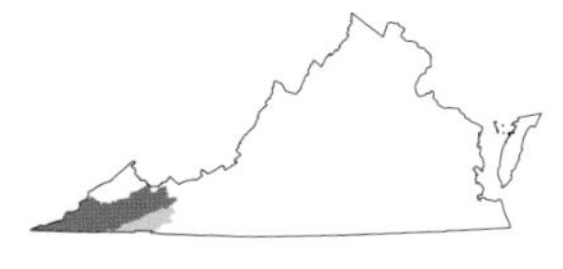

■ Bluebreast Darter
■ Greenfin Darter

Description: Rounded frontal view, elongate profile; moderately short snout; blue breast; brown to olive above and on side with dark, narrow horizontal stripes, brown to tan below; breeding males with red (Bluebreast Darter) to orange (Greenfin Darter) spots on side; fin color dusky olive to orange (Bluebreast Darter) or olive to brilliant green (Greenfin Darter); white-and-gold outline with dark margins on second dorsal, caudal, and anal fins. **Reproduction:** Spawn in late spring and early summer; bury eggs in sand or fine gravel. **Food:** Aquatic invertebrates. **Notes:** These species are very similar in appearance but occur in different drainages. Greenfin Darter is listed as threatened in Virginia, where it is nearly confined to a single stream system.

Redline Darter

Etheostoma rufilineatum
Syn.: *Nothonotus rufilineatus*

Size: 2–3 in.
Habitat: Warm medium to large streams and rivers.
Abundance: Common.
Status: Native.

Description: Slightly compressed frontal view, moderate profile; pointed snout; tan to brown above, "plaid" pattern of brown, red, and black on side (in male only), tan to white below; cream-colored hourglass at base of caudal fin; fins of male clear to dusky

Redline Darter
continued:

with red margins; fins of female typically clear with many black spots; often blue or turquoise breast and spot behind gills. Reproduction: Spawns in spring and summer; buries eggs in gravel of riffles. Food: Aquatic invertebrates. Notes: Presence of the Redline Darter is a good indicator of healthy riffle habitat, as this species is highly dependent on clean rocky areas with fast current. It was recently introduced in the New River system and continues to expand its range in the drainage.

Golden Darter

Etheostoma denoncourti
Syn.: *Nothonotus denoncourti*

male

Size: ¾–1 in.
Habitat: Warm medium to large rivers.
Abundance: Rare.
Status: Native.

Description: Compressed frontal view, moderate profile; moderately pointed snout; medium brown with dark-brown speckling (female) to pale yellow with orange and brown speckling (nonbreeding male) above, on side, and below; dark bar around caudal peduncle with two faint pale spots on caudal-fin base; dark saddle behind head; breeding male orange yellow to gold with multiple bars on side; dark-blue throat; fins clear or slightly dusky (female) to lemon yellow (male). Reproduction: Spawns in summer; buries eggs in sand-gravel. Food: Aquatic invertebrates. Notes: Golden Darter is one of the smallest darter species. In Virginia, it is listed as threatened and only known from the mainstem Clinch River.

Sharphead Darter

Etheostoma acuticeps
Syn.: *Nothonotus acuticeps*

Size: 1–2½ in.
Habitat: Warm medium to large rivers.
Abundance: Rare.
Status: Native.

Description: Moderately compressed frontal view, moderate profile; extremely pointed snout; female and nonbreeding male yellow brown above and on side, with light dusky bars, paler below; breeding male olive to dark blue green, dark vertical bars on side, paler below; fins dusky in female to turquoise in breeding male. Reproduction: Spawns in summer; buries eggs in gravel. Food: Aquatic invertebrates. Notes: Sharphead Darter is extremely rare in Virginia and listed as endangered. The only known population occurs in the Holston River system.

Wounded Darter

Etheostoma vulneratum
Syn.: *Nothonotus vulneratus*

Size: 1½–2½ in.
Habitat: Warm medium to large rivers.
Abundance: Uncommon.
Status: Native.

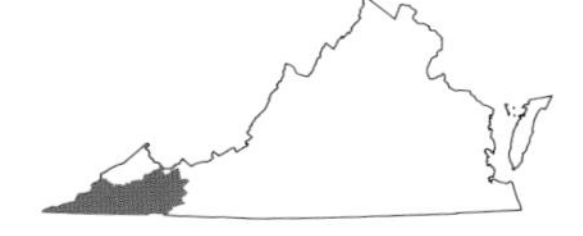

Description: Rounded frontal view, moderate profile; pointed snout; pale stripes on side, darker posteriorly; dark olive above, light olive to gray on sides, with bright red spots, gray to pale green below; fins slightly dusky to dark gray, sometimes slightly green; male has red spots on first dorsal fin, red tint on caudal fin. Reproduction: Spawns in late spring; attaches eggs to the underside of rocks. Male guards eggs until hatching. Food: Aquatic invertebrates. Notes: Wounded Darter is named for the red spots that resemble puncture wounds on the breeding male. They tend to be secretive and are seldom seen.

Rainbow Darter

Etheostoma caeruleum

male

Size: 1–3 in.
Habitat: Warm small streams to large rivers.
Abundance: Uncommon.
Status: Native.

Description: Strongly compressed frontal view, moderate profile; moderately rounded snout; 6–10 short dark saddles; female brown, tan, or pale yellow above and on side, white below; dark mottling above and small blotches on side; breeding male with blue to blue-green diagonal bars, interspersed with red-orange areas transitioning to mostly orange near caudal fin; orange throat and blue cheek; orange-and-blue bands on most fins except pectoral and pelvic fins; female fin coloration appears very faint. Reproduction: Spawns from spring to early summer; buries eggs in fine gravel. Food: Aquatic invertebrates. Notes: Rainbow Darter is a beautiful and popular aquarium specimen. It is currently spreading upstream in the New River drainages.

Spawning female and male Candy Darter.

Fantail Darter

Etheostoma flabellare (right)

Carolina Fantail Darter

Etheostoma brevispinum (not shown)

Size: 1–2½ in.
Habitat: Cool to warm small streams to medium rivers.
Abundance: Common.
Status: Native.

Description: Rounded frontal view, slightly elongate profile; snout slightly pointed (Fantail Darter) to slightly blunt (Carolina Fantail Darter); tan to brown (often orange in Carolina Fantail Darter) above and on side, tan to white below; 7 or fewer (Carolina Fantail Darter) or 8–10 (Fantail Darter) dark bars on side, extending to back; bars less prominent in females and juveniles; fins clear to yellowish with dark lines, especially on caudal fin; yellow to gold knobs at tips of dorsal spines of breeding males. Reproduction: Spawn in spring; lay eggs on the underside of flat rocks; males guard eggs until hatching. Food: Aquatic invertebrates. Notes: These generalists are often the most common darters in small streams and are tolerant of a wide variety of habitats.

Duskytail Darter

Etheostoma percnurum (not shown)

Size: 1–2 in.
Habitat: Warm medium streams and rivers.
Abundance: Rare.
Status: Native.

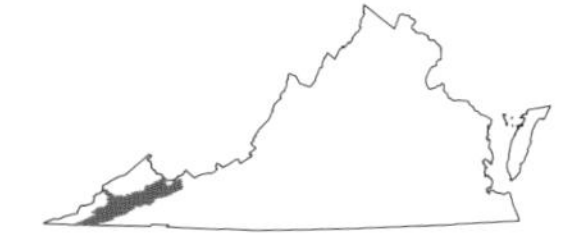

Description: Rounded frontal view, slightly elongate profile; dusky brown to gray above, yellow to light brown on side, dusky gray to opaque white below; small dark saddles connect to 11–14 dark narrow bars on side; fins clear to slightly yellow with orange or yellow knobs at tips of anterior dorsal-fin spines; pectoral, anal, and caudal fins with dark dusky margins; colors in breeding male intensify. Reproduction: Spawns in spring; lays eggs on the underside of flat rocks; male guards nest site. Food: Aquatic invertebrates. Notes: Duskytail Darter has one of the smallest ranges of any fish species and is protected federally by the Endangered Species Act. Efforts are under way to help recover this species via captive propagation.

Swamp Darter

Etheostoma fusiforme (right)

Carolina Darter

Etheostoma collis
(not shown)

Sawcheek Darter

Etheostoma serrifer (not shown)

Size: 1–2¾ in.
Habitat: Warm small streams, ponds, and swamps.
Abundance: Rare to common.
Status: Native.

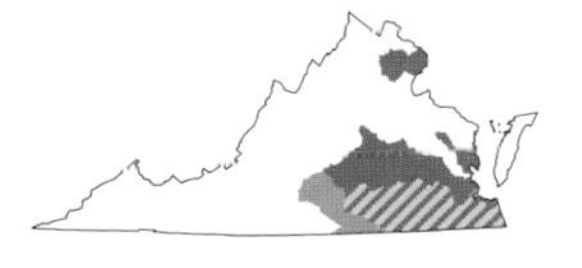

- Swamp Darter
- Carolina Darter
- Sawcheek Darter

Description: Slightly compressed frontal view, elongate profile; short snout; tan, brown, or olive above and on side with dark blotches, sometimes with green iridescence, tan to pale olive below; fins transparent and tessellated; one (Carolina Darter and Swamp Darter) to two (Sawcheek Darter) distinct black spots in middle of caudal-fin base. **Reproduction:** Spawn in spring; attach eggs to debris or vegetation. **Food:** Aquatic invertebrates. **Notes:** These drabbly colored darters prefer slow or still waters, which are avoided by most darter species. The Carolina Darter is listed as a threatened species in Virginia.

Male Riverweed Darter in spawning color.

DRUMS
Family Sciaenidae

Members of the family Sciaenidae are found worldwide, predominately in marine environments. It includes notable marine sporting species such as Red Drum, sea trouts, croakers, and kingfish; however, only Freshwater Drum inhabits freshwater habitats in North America. Distinguishable external characteristics are a deep body, highly arched back, two dorsal fins (short first and long second), one to two anal spines, and a lateral line. Freshwater Drum inhabits lakes, reservoirs, and large rivers. Its name derives from the ability to make a drumming sound using its gas bladder as a resonating chamber.

Freshwater Drum
Aplodinotus grunniens

Size: 10 in. to 3 ft.
Habitat: Warm large rivers and reservoirs.
Abundance: Rare.
Status: Native.
Other name: Gaspergou.

Description: Strongly compressed frontal view, deep profile; two dorsal fins; steeply sloped between dorsal fin and head; short head; subterminal mouth; dark above, dusky silver on sides, pale below; pelvic fins light orange. **Reproduction:** Spawns in spring and summer; female may lay between 27,000 and 508,000 eggs. **Food:** Aquatic insects, crustaceans, snails, and mussels. **Notes:** Freshwater Drum is native to the Tennessee River drainage and introduced to Kerr Reservoir, where the current state record of 26 lb., 8 oz. was caught.

SNAKEHEADS
Family Channidae

Among the 51 species of snakeheads worldwide, only 1 species, the Northern Snakehead, *Channa argus*, an endemic to China, Russia, and Korea, is known to inhabit the waters of the Commonwealth. The Northern Snakehead was first discovered in the tidal Potomac River in 2004 and has since gained national attention due to its unusual appearance, piscivorous nature, and negative media hype. It was established in the Rappahannock River basin and Piankatank River by 2016 and in the James River system in 2018. It is similar in appearance to our native Bowfin, and its young are often confused with darters. The Northern Snakehead can survive out of water for long periods because it is an obligate air breather, but it does not migrate over dry land. It overwinters by burrowing in the mud. It has been extensively studied in tidal Potomac River tributaries of Virginia, and its distribution is being closely monitored by the state's fisheries biologists.

Northern Snakehead
Channa argus

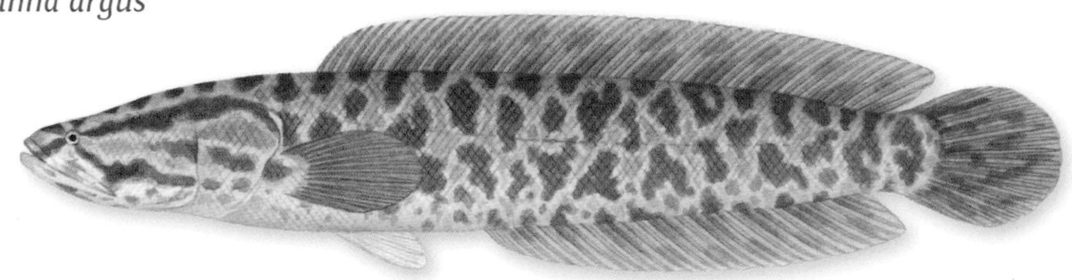

Size: 20 in. to 3 ft.
Habitat: Tidal creeks, rivers, freshwater lakes, and ponds.
Abundance: Uncommon.
Status: Introduced.

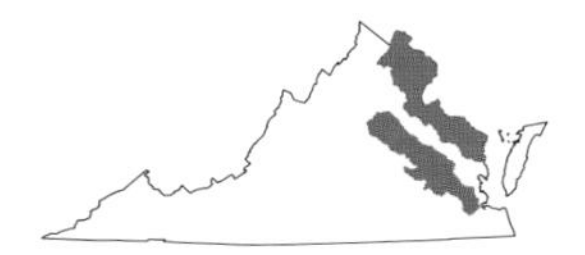

Description: Rounded frontal view, elongate profile; terminal mouth with many sharp teeth; scaled head; brown body with black mottled blotches on sides; long single dorsal and anal fins. Reproduction: Spawns in spring or early summer; makes pre-spawn migration; newly hatched young form schools in the shape of large balls and are protected by both parents. Food: Small fishes, particularly Banded Killifish and Bluegill. Notes: Northern Snakehead is highly sought after by certain anglers for its fighting ability and flavor.

GLOSSARY

Abdominal: Located on the ventral surface of the body between pelvic and anal fins.
Adipose fin: A fleshy fin, without supporting rays, behind the dorsal fin.
Ammocoete: Non-parasitic lamprey larvae.
Anadromous: Migrating from marine waters into fresh waters to spawn.
Anal fin: The fin on the underside of the body, typically just behind the anus.
Apex predator: Predator, at the top of a food chain, with few natural predators.
Atlantic Slope basin: Refers to drainages that flow into the bays and estuaries of the Atlantic Ocean.
Band: A broad, pigmented, vertical or diagonal line.
Barbel: A whisker-like projection of fleshy tissue, usually found around the mouth, chin, or nostrils.
Base: Part of fin joined to the body.
Benthic: Living on or associated with the bottom of a water body.
Brackish: Slightly salty water, due to mixture of fresh water and seawater.
Carnivore: An animal that preys and feeds on other animals.
Catadromous: Migrating from fresh waters to marine waters to spawn.
Caudal: Pertaining to the tail.
Caudal fin: Fin at the tail end of a fish.
Cohort: Fish of the same species born during the same year.
Compressed: Flattened from side to side in cross section and higher than wide.
Ctenoid scale: Thin scales with small spines on the exposed edge.
Cycloid scale: Thin smooth scales without spines on the exposed edge.
Detritus: Bottom material generated by water and decomposing materials.
Dorsal: Pertaining to the back or upper surface of the body.
Dorsum: The dorsal surface of the body.
Dorsal fin: Fin on the midline of the back; may be notched or divided.
Drainage: A group of interconnected streams and rivers that empty into the ocean, an estuary, or the mainstem of a basin.
Elongate: In fishes, describes a body length proportionally greater than body depth.
Emarginate: Describes a caudal fin that is notched but not deeply forked.
Endemic: Native or restricted to a certain area.
Epilimnion: The warmer, upper layer of stratified lake, reservoir, or pond water.
Estuary: Semi-enclosed tidal body of water of variable salinity with a free connection to the sea.
Extirpated: Eliminated from a particular region; locally extinct.
Falcate: Strongly inwardly curved.
Family: A taxonomic level that ranks below order and may contain one or more genera.
Fecund: Describes a female fish carrying large numbers of mature eggs.

Filament: A long, thread-like structure.
Filter feeder: A fish that feeds by sifting plankton from the water.
Frenum: A bridge of tissue connecting the upper jaws to the snout.
Fusiform: Spindle-shaped and tapered at both ends.
Ganoid scale: Scale type that is thick, hard, and diamond-shaped.
Generalized: Refers to fishes with a basic or ancestral trait.
Genus (plural genera): A principal taxonomic category that ranks above the species and below the family.
Gill: A respiratory organ that enables aquatic animals to take oxygen from water and to excrete carbon dioxide.
Gill arch: Bony or cartilaginous structure bearing gill filaments and gill rakers, located inside the gill chamber.
Gill cover: Also known as the operculum.
Gill opening: Exit where water used in respiration leaves the body of the fish. May be pore-like, as in lampreys, or slit-like.
Gill rakers: Slender, bony or hardened projections along the inner side of a gill arch.
Glochidia: The unmetamorphosed larvae of a freshwater mussel.
Grow-out: Act of growing fishes beyond useful size to larger, more desirable size for recreational opportunities or commercial markets.
Gular plate: A bony plate on the throat between the lower jaw bones, present in some fishes.
Heterocercal: Caudal fin or tail with larger upper lobe than lower and with the vertebral column extending into the upper lobe.
Humeral: Area just behind the gill opening and above the pectoral-fin base.
Hybrid: Offspring of two distinct species.
Inferior: Located on the underside of the head, with the upper jaw projecting beyond the lower jaw.
Insertion: The place where the posterior-most point of a fin base is joined to the body.
Intergrade zone: Area inhabited by fishes with intermediate characteristics of multiple parent species or subspecies due to hybridization.
Introduced: Describes fishes relocated to an area from outside the native or indigenous range.
Invertivore: An animal that primarily feeds on invertebrates.
Keel: In fishes, a sharp, fleshy, or bony ridge.
Kype: Hook-shaped lower jaw on an adult male trout or salmon.
Larva (plural larvae): Early developmental stage following the emergence from the egg and preceding the transformation to the juvenile stage.
Lateral: At, pertaining to, or in the direction of the side of a fish.
Lateral line: A connected series of tubes or pores associated with the sensory canal system that detects movement in the water.
Leptocephalus (plural leptocephali): Translucent, ribbon-like pelagic larvae.
Median fins: The unpaired fins originating from the midline of a fish, specifically the dorsal, anal, and caudal fins.

Mottling: Blotchy or marbled pattern.
Mound: A pile of gravel and pebbles constructed by a fish for the purpose of spawning and incubating eggs.
Naturalized: Describes an introduced species that has assimilated into a native fish community with limited observed negative effects or provides benefits to the aquatic community and society that outweigh negative effects.
Nest associate: Members of family Cyprinidae that spawn exclusively or opportunistically on a chub mound.
Nonnative species: Not originating or occurring naturally in a particular place.
Nuptial: Characteristic of or occurring in the breeding season.
Oblique: Neither perpendicular nor parallel; at an angle.
Ocellus (plural ocelli): A pigmented eye-like spot, usually dark with a lighter border.
Omnivore: An animal that feeds on both plants and animals.
Opercle: The uppermost and largest of the bones that form the operculum. Adjective: opercular.
Operculum: A movable multi-bone plate covering the gill chamber; also known as the gill cover.
Otolith: Each of three small calcareous bodies in the inner ear of some vertebrates, involved in sensing vibrations (hearing), orientation, and movement.
Papilla: a small rounded protuberance on a part or organ of the body.
Papillose: Refers to small, round or cone-shaped bumps on the lips of fishes.
Parr: Life stage of a young salmon or trout following the fry stage.
Parr marks: Dark round patches on sides of trouts and salmons in the parr life stage.
Pectoral fin: One of the paired fins located behind the gill opening.
Pelagic: Inhabiting the middle layers of an open body of water.
Pelvic fin: Paired fins on the lower part of the body.
Pharyngeal: Relating to the throat region.
Piscivore: Animal that feeds mostly on fishes.
Plicate: Folded, grooved, or wrinkled.
Posterior: In fishes, located toward the rear.
Premaxillary: Pertaining to the anterior and usually smaller of the two bones forming the upper jaw.
Preopercle: The anterior-most bone of the operculum, below and behind the eye.
Ray: One of the supporting structures in the fin membrane; may be either stiff (spine) or flexible (soft ray).
Recruitment: Number of fishes surviving to enter a fishery or an age or life history stage of importance.
Redd: Gravel nest usually associated with trouts and salmons.
Riffle: A rocky or shallow part of a stream or river with turbulent water.
River basin: Portion of land drained by a river and its tributaries.
Run: Habitat intermediate of a pool and riffle in a river or stream.
Saddle: A blotch or patch of pigment extending from the midline of the back onto the sides.

Scale: In fishes, one of the many hard or bony plates that cover the skin.
Scute: A modified scale, often large and shield-like, with one or more keel-like ridges.
Sensory pore: A small opening leading to the lateral line canal system on the head and/or body.
Sexual dimorphism: Difference in size or appearance of sexes of a species.
Snout: The part of the head in front of the eye.
Spawn: To release eggs and sperm, usually into water by aquatic animals, including fishes.
Specialized: Highly differentiated from a basic generalized form.
Species: A group of interbreeding populations that are reproductively isolated from other such groups; a subset of a genus.
Spine: An unbranched and usually rigid, unsegmented fin ray; also a sharp bony projection on the head or body.
Stippling: Numerous small dots or specks.
Stock: A management unit with homogenous rates of growth, mortality, and recruitment of interest to fishery managers.
Subterminal: A mouth where the upper jaw slightly projects beyond the lower and the snout extends in front of the mouth.
Superior: A mouth in which the lower jaw projects beyond the upper.
Swim bladder: A gas- or fat-filled sac that provides buoyancy.
Tailout: Habitat that leads away from an adjacent habitat, e.g., pool tailout.
Taxonomic: Concerned with the classification of organisms, such as a species, family, or class.
Tear drop: A subocular bar; pigment patch below the eye.
Terminal: A mouth with the anterior-most points of the upper and lower jaws extending to the same place.
Tessellated: Having a checkered or repeated pattern.
Thermocline: The middle layer of a body of water, noted for dramatic change in temperature.
Thoracic: Describes a pelvic fin located on the breast area (as opposed to the belly), below or ahead of the pectoral fin.
Tooth patch: An isolated patch of teeth; usually on the roof of the mouth or the tongue.
Triploid: Possessing three homologous sets of chromosomes.
Truncate: Describes a caudal fin with a squared-off rear margin.
Tubercles: Small bumps, sometimes pointed, usually found on male fishes during breeding.
Undescribed: A taxon (e.g., species) that has been discovered but not yet formally described and named.
Vermiculation: Markings similar to the pattern made by a worm boring through wood.
Vomer teeth: Teeth on the middle of the roof of the mouth.
Watershed: An area of land that drains water to the same river, basin, or sea.
Winterkill: Fish deaths associated with low oxygen concentrations during the winter.
Year-class: All fishes of the same species born in the same year.

SELECT BIBLIOGRAPHY

The complete bibliography can be found online at: www.press.jhu.edu. In the search box, type: "Field Guide to Freshwater Fishes of Virginia."

Acre, H.M., J.G. Lundberg, and M.A. O'Leary. 2017. Phylogeny of the North American catfish family Ictaluridae (Teleostei: Siluriformes) combining morphology, genes and fossils. *Cladistics* 33:406–428.

Alford, J.B., and D.C. Beckett. 2006. Dietary specialization by the Speckled Darter, *Etheostoma stigmaeum*, on chironomid larvae in a Mississippi stream. *Journal of Freshwater Ecology* 21:543–551.

Angermeier, P.L., and M.J. Pinder. 2015. Viewing the status of Virginia's environment through the lens of freshwater fishes. *Virginia Journal of Science* 66:147–169.

Blanton, R.E., and R.E. Jenkins. 2008. Three new darter species of the *Etheostoma percnurum* species complex (Percidae, subgenus *Catonotus*) from the Tennessee and Cumberland river drainages. *Zootaxa* 1963:1–24.

Blanton, R.E., and G.A. Schuster. 2008. Taxonomic status of *Etheostoma brevispinum*, the Carolina fantail darter (Percidae: *Catonotus*). *Copeia* 2008(4):844–857.

Boschung, H.T., and R.L. Mayden. 2004. *Fishes of Alabama*. Washington, DC: Smithsonian Press.

Fricke, R., W.N. Eschmeyer, and R. van der Laan, editors. 2018. Catalog of fishes: Genera, species, references. http://researcharchive.calacademy.org/research/ichthyology/catalog/fishcatmain.asp.

Goldstein, R.J. 2000. *American Aquarium Fishes*. College Station: Texas A&M University Press.

Hilling, C.D., S.L Wolf, J.R. Copeland, D.J. Orth, and E.M. Hallerman. 2018. Occurrence of two non-indigenous catostomid fishes in the New River, Virginia. *Northeastern Naturalist* 25:215–221.

Jelks, H.L., et al. 2011. Conservation status of imperiled North American freshwater and diadromous fishes. *Fisheries* 33(8):372–407.

Jenkins, R.E., and N.M. Burkhead. 1994. *Freshwater Fishes of Virginia*. Bethesda: American Fisheries Society.

Kansas Fishes Committee. 2014. *Kansas Fishes*. Lawrence: University Press of Kansas.

Keck, B.P., and T.J. Near. 2008. Assessing phylogenetic resolution among mitochondrial, nuclear, and morphological datasets in *Nothonotus* darters (Teleostei: Percidae). *Molecular Phylogenetics and Evolution* 46:708–720.

Kestemont, P., K. Dabrowski, and R.C. Summerfelt, editors. 2015. Biology and culture of percid fishes: Principles and practices. New York: Springer.

Kinzinger, A.P., R.L. Raesly, and D.A. Neely. 2000. New species of *Cottus* (Teleostei: Cottidae) from the Middle Atlantic Eastern United States. *Copeia* 2000:1007–1018.

Kinzinger, A.P., R.M. Wood, and D.A. Neely. 2005. Molecular systematics of the genus *Cottus* (Scorpaeniformes: Cottidae). *Copeia* 2005:303–311.

Kraczkowski, M.L., and B. Chernoff. 2014. Molecular phylogenetics of the eastern and western blacknose dace, *Rhinichthys atratulus* and *R. obtusus* (Teleostei: Cyprinidae). *Copeia* 2014:325–338.

Layman, S.R., and R.L. Mayden. 2012. Morphological diversity and phylogenetics of the darter subgenus *Doration* (Percidae: *Etheostoma*), with descriptions of five new species. *Bulletin of the Alabama Museum of Natural History* 30:1–83.

Murdy, E.O., J.A. Musick, and V. Kells. 2013. *Field Guide to Fishes of the Chesapeake Bay*. Baltimore: Johns Hopkins University Press.

Near, T.J., E.D. France, B.P. Keck, and R.C. Harrington. 2016. Systematics and taxonomy of the Snubnose Darter, *Etheostoma simoterum* (Cope). *Bulletin of the Peabody Museum of Natural History* 57:127–145.

Page, L.M., and B.M. Burr. 2011. *Peterson Field Guide to Freshwater Fishes of North America North of Mexico*. New York: Houghton Mifflin Harcourt Publishing.

Page, L.M., et al. 2013. *Common and Scientific Names of Fishes from the United States, Canada, and Mexico*. 7th edition. Bethesda: American Fisheries Society, Special Publication 34.

Robins, C.R. 2005. *Cottus kanawhae*, a new cottid fish from the New River System of Virginia and West Virginia. *Zootaxa* 987:1–6.

Robins, R.H., L.M. Page, J.D. Williams, Z.S. Randall, and G.E. Sheehy. 2018. *Fishes in the Fresh Waters of Florida: An Identification Guide and Atlas*. Gainesville: University of Florida Press.

Rohde, F.C., R.G. Arndt, J.W. Foltz, and J.M. Quattro. 2009. *Freshwater Fishes of South Carolina*. Columbia: University of South Carolina Press.

Simon, T.P. 2011. *Fishes of Indiana: A Field Guide*. Bloomington: University of Indiana Press.

Stauffer, J.R., Jr., R.W. Criswell, and D.P. Fischer. 2016. *The Fishes of Pennsylvania*. El Paso: Cichlid Press.

Tan, M., and J.W. Armbruster. 2018. Phylogenetic classification of extant genera of fishes of the order Cypriniformes (Teleostei: Ostariophysi). *Zootaxa* 4476(1):6–39.

U.S. Department of the Interior, U.S. Fish and Wildlife Service, and U.S. Department of Commerce, U.S. Census Bureau. 2016 National Survey of Fishing, Hunting, and Wildlife-Associated Recreation.

Warren, M.L., Jr. 2009. Centrarchid identification and natural history. Pages 375–533 in S.J. Cooke and D.P. Philipp, editors. Centrarchid Fishes: Diversity, Biology, and Conservation. United Kingdom: Blackwell Publishing, Ltd.

Warren, M.L., Jr., and B.M. Burr. 2014. *Freshwater Fishes of North America*. Volume 1. Petromyzontidae to Catostomidae. Baltimore: Johns Hopkins University Press.

Websites with useful resources and information about fishes:

Fishbase: http://fishbase.org/home.htm
Florida Museum of Natural History, Ichthyology:
https://www.floridamuseum.ufl.edu/fish/
IUCN Red List of Threatened Species: https://www.iucnredlist.org/
North American Native Fishes Association (NANFA): http://www.nanfa.org/
Smithsonian National Museum of Natural History, The Division of Fishes:
http://vertebrates.si.edu/fishes/index.html

Periodical of interest:

American Currents, published quarterly by the NANFA: http://www.nanfa.org/

SCIENTIFIC NAME INDEX

COMMON NAME INDEX

CREDITS

Illustration credits:

Val Kells: pp. 12, 43, 46 bottom (b), 48, 52, 53, 55, 56, 57, 58, 59, 61, 62, 63, 64, 65, 66 top (t), 68 (t), 69 (b), 70 (t), 74 (b), 75, 76 (b), 78, 80, 81 (t), 84, 86, 87 (t), 88 (b), 91, 92 (b), 93 (b), 94, 95 (t), 96 (t), 98 (b), 101 (b), 102 (b), 103, 104, 105, 106 (t), 107 (b), 112, 113 (b), 114, 115, 116, 117, 119, 120 (t), 121, 123, 124, 130, 131, 133, 136, 137, 138, 139, 140, 141, 143, 144 (b), 146 (t), 147, 150, 153, 154 (b), 155 (b), 156, 157, 160 (b), 161 (b), 162, 166, 167, 169, 170, 175

Joseph R. Tomelleri: pp. 45, 46 (t), 50, 51, 66 (b), 67, 68 (b), 69 (t), 70 (b), 72, 73, 74 (t), 76 (t), 77, 79, 82, 83, 85, 87 (b), 88 (t), 89, 92 (t), 93 (t), 95 (b), 96 (b), 97, 98 (t), 101 (t), 102 (t), 106 (b), 107 (t), 108, 110, 111, 113 (t), 118, 120 (b), 126, 127, 128, 134, 135, 142, 144 (t), 146 (b), 149, 151, 152, 154 (t), 155 (t), 158, 159, 160 (t), 161 (t), 163, 164, 165, 168, 171, 172, 173, 174

Photo credits:

Brian Borkholder: p. 148, Smallmouth Bass sac fry
Paul E. Bugas, Jr.: p. 29, Smallmouth Bass with attached radio tag
Louise Finger: p. 31, before and after stream restoration
Sean Landsman: p. 49, juvenile Atlantic Sturgeon
Tim Lane: p. 22, Freshwater mussel, *Lampsilis fasciola*, darter-like lure
Christine Lisle: p. 24, underwater photographer
Meghan Marchetti: p. 22, young angler fishing
Donald J. Orth: p. 25, researchers with seine
Michael J. Pinder: p. 18, Mountain stream riffle, Piedmont stream run, Ridge and Valley pool, Coastal Plain millpond
Michael St. Germain: p. 29, biologists electrofishing and using seine to sample fishes
Isaac Szabo: p. 99, underwater Golden Redhorse; p. 122, underwater Northern Studfish
Derek A. Wheaton: p. 23, Saffron and Tennessee Shiners on chub spawning mound, ichthyology students snorkeling; p. 24, underwater Saffron and Tennessee shiners; p. 25, Bluespotted Sunfish in photo tank; p. 27, Greenside Darter in aquarium, Dollar Sunfish in aquarium; p. 78, underwater White Shiner, Central Stoneroller, Crescent Shiner, and Mountain Redbelly Dace; p. 81, Tennessee Shiners in breeding coloration; p. 111, Brook Trout; p. 114, underwater Chain Pickerel; p. 164, Blueside Darter spawning pair; p. 171, Candy Darter spawning pair; p. 173, Riverweed Darter

ABOUT THE AUTHORS
AND ILLUSTRATORS

Paul E. Bugas, Jr., is a Fisheries Manager with the Virginia Department of Game and Inland Fisheries based in Verona. He graduated from Virginia Tech in 1975 with a Bachelor of Science degree in forestry and wildlife (fisheries option), immediately starting his career with VDGIF. He currently oversees fish stock assessments, habitat improvement, fish stocking, environmental impacts, fish health, and water quality projects in 29 counties. Bugas has received awards from the American Fisheries Society, the Garden Club of Virginia, Trout Unlimited, the Izaak Walton League of America, and others for his outreach efforts in aquatic education. In 1990, Paul organized the Virginia Chapter of the American Fisheries Society and served as its first president.

Corbin D. Hilling is a Virginia Sea Grant Graduate Research Fellow and PhD student at Virginia Tech. Corbin completed degrees in biology (BS) and wildlife and fisheries resources (MS) at West Virginia University. He is an award-winning instructor in the Virginia Tech Department of Fish & Wildlife Conservation, where he teaches students to identify Virginia fishes. Corbin has published several scientific articles on fishes and their biology during his young career. He is active in public outreach and engagement to spread awareness of fishes and their conservation. During his free time, Corbin enjoys fishing for both freshwater and marine fishes.

Val Kells is a renowned Marine Science Illustrator who works with designers, curators, biologists, and master planners to create illustrations for educational graphic displays. Her work has appeared in over 30 public aquariums, museums, and nature centers. Val also works with publishers and authors to create a wide variety of books and periodicals. She most recently co-authored and illustrated *A Field Guide to Coastal Fishes: From Maine to Texas* and *A Field Guide to Coastal Fishes: From Alaska to California* and she is currently working on the third in that series. Val also illustrated *Field Guide to Fishes of the Chesapeake Bay* and *Tunas and Billfishes of the World*, all published by Johns Hopkins University Press. Val is an avid fisherman and naturalist and in her off time, she loves to fish and explore with her two sons.

ABOUT THE AUTHORS
AND ILLUSTRATORS

Michael J. Pinder is a Frostburg State University graduate with degrees in biology (BS) and fisheries management (MS). He has spent the last 25 years as an aquatic biologist for the Virginia Department of Game and Inland Fisheries. Mike has worked on a variety of taxa groups including freshwater mollusks, fishes, amphibians, and reptiles. He established the agency's first facility to propagate and recover freshwater mussels of the Tennessee River drainage. His primary focus is the management and conservation of Virginia's nongame and endangered freshwater fishes. He has been honored with multiple awards, the most recent being the Thomas Jefferson Award for Conservation. Mike is a lifelong angler, naturalist, woodworker, and artist.

Derek A. Wheaton is a biologist for Conservation Fisheries, Inc., which breeds and monitors rare fish species for conservation efforts. He has worked for the Virginia Department of Game and Inland Fisheries, performing statewide surveys for rare and nongame fish species. He is a regional representative and board member for the North American Native Fishes Association. He is an avid underwater photographer, spending most of his free time in a river or stream and sharing photos of amazing aquatic wildlife as Enchanting Ectotherms on social media. He is an accomplished aquarist, specifically with native species, and has maintained many of Virginia's fishes in captivity.

Donald J. Orth is the Thomas H. Jones Professor in the Department of Fish and Wildlife Conservation at Virginia Polytechnic Institute and State University. Don attended Eastern Illinois University (BS) and Oklahoma State University (MS and PhD). He is a Life Member of the American Fisheries Society and a Certified Fisheries Professional. He is also a Fellow of the American Fisheries Society, the American Institute of Fisheries Research Biologists, and the Virginia Natural Resources Leadership Institute. In addition to over 100 popular writings, Don has published over 200 technical works on fishes, fisheries, and riverine management and has received numerous awards for his teaching and contributions to conservation.

ABOUT THE AUTHORS
AND ILLUSTRATORS

Joseph R. Tomelleri is a naturalist, biologist, avid fisherman, and artist. He uses Prismacolor, graphite pencils, and attention to precise detail learned from personal collecting and photographing native fishes. His 1,100+ illustrations have appeared in numerous publications, including *Trout and Salmon of North America*, *Fishes of Alabama*, *Fishes of Kansas*, *Fishes of Indiana*, *Fishes of Texas*, and *Freshwater Fishes of North America*. Recent projects include *Native Trout of Mexico*, *Fishes of Idaho*, and *Fishes of the Salish Sea and Puget Sound*. He holds a BS and MS in biology from Fort Hays State University. He lives with his wife, Susan, and their two sons in Leawood, Kansas.

Also from Johns Hopkins University Press:

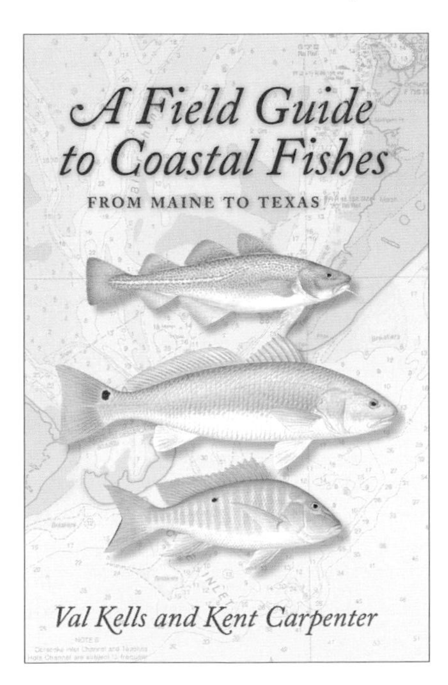

ISBN 13: 978-0-8018-9838-9
ISBN 10: 0-8018-9838-2

ISBN 13: 978-1-4214-0768-5
ISBN 10: 1-4214-0768-X

NOTES

NOTES

NOTES